A Great Mystery Unraveled:

The Secret of WHY the Speed of Light Is Constant Finally Unveiled

ALSO BY BINGCHENG ZHAO

(Popular Science & Science)

Why It's Difficult to Understand "A Brief History of Time"

Terminate the Controversy over the Big Bang Theory by Inspecting All Its Three Pillars

Dark Matter Is No Longer in the Darkness! (The Constituents of Dark Matter Have Been Revealed)

$E = mc^2$ *Talks with the Law of Mass Doing Work*

Look at the Mechanism behind the Postulate of the Equivalence Principle

From Postulate-Based Modern Physics to Mechanism-Revealed Physics

(Science & Religion)

The Amazing Wisdom and Truth: GOD Can Change the Scales of Space and Time in the Universe

The Naked Truth: The Big Bang Theory Cannot Deny GOD!

The Hard Evidence: The Big Bang Theory Has No Way to Deny GOD! (After Inspecting All the Three Pillars of the Big Bang Theory)

A Great Mystery Unraveled:

The Secret of WHY the Speed of Light Is Constant Finally Unveiled

Bingcheng Zhao

THE SECRET OF WHY THE SPEED OF LIGHT IS CONSTANT
FINALLY UNVEILED

Published in the United States of America through Amazon's KDP

ISBN-13: 978-1694414991

CONTENTS

FOREWORD

A lot of people, who have been baffling over the question of *why* the speed of light is constant, wish they would no longer be baffled by this question. A lot of people, who have been thinking over the question of *why* the speed of light is constant, hope to know the answer to this question. A lot of people, who have known the speed of light is constant, want to know the mechanism or reason *why* the speed of light is constant.

For more than one century, the long-standing, big question of *why* the speed of light is constant has constantly baffled human beings. And we human beings will continue to be baffled by this long-standing, big question if it remains unanswered. Therefore, we human beings must find out the answer to this long-standing, big question by revealing the mechanism or reason *why* the speed of light is constant!

Unfortunately or regrettably, also unavoidably or inevitably, we human beings haven't answered, cannot answer, and will never answer the question of *why* the speed of light is constant within the paradigms of (postulate-based) special relativity and (postulate-based) modern physics. WHY? The answer is: because 'the speed of light is constant' is one of the two <u>postulates</u> of special relativity that is not only an important component of modern physics, but also an indispensable foundation of modern physics (that is to say, the question of *why* the speed of light is constant was created from special relativity; special relativity and modern physics are on the same level of consciousness);

because, according to Albert Einstein, "no problem can be solved from the same level of consciousness that created it."

Fortunately, as presented in this book, the answer to the question of *why* the speed of light is constant has been found out, because the mechanism or reason *why* the speed of light is constant has been revealed; that is, the cause of *why* the speed of light is constant has been brought to light. And so, from now on, people will no longer continue to be baffled by the question of *why* the speed of light is constant if they want to know the answer to this question.

As shown in this book, the main route along which the mechanism or reason *why* the speed of light is constant has been revealed is: the law of mass doing work (which is a newly discovered and verified physical law about *why* mass has energy) → the mass consumption (a direct and inevitable product of the law of mass doing work) → the newly developed and verified *mechanism-revealed scales relativity theory* (this new theory, being the result of an application of the mass consumption thus being the product of the law of mass doing work, is the *only* scientific theory that is able to tell us *why* time runs slower at high speed. Related clarification or reminder: the theory of special relativity, because it is unable to know *why* the speed of light is constant, doesn't have the ability to know *why* time runs slower at high speed) → the mechanism or reason *why* the speed of light is constant. (Narrator: this main route indicates that this book is about new concepts, new discoveries and new knowledge, though the core target of this book is about solving the 'age-old' problem or answering the 'age-old' question, which is *why* the speed of light is constant.)

This main route shows that the law of mass doing work is the starting point, prerequisite and key to revealing the mechanism or reason *why* the speed of light is constant; that is, only this law has been discovered, can we human beings find out the mechanism or reason *why* the speed of light is constant. (Commentator: and so, the long-standing question of *why* the speed of light is constant is anxiously waiting for the law of mass doing work; all the people, who have been yearning to know *why* the speed of light is constant, are eagerly looking forward to knowing this law.)

Because the four components on the main route above reflect and represent the general framework of this book, let us concisely preview them here.

The law of mass doing work is the *only* physical law that answers the fundamental question of *why* mass has energy by revealing and showing that mass has the capability of doing work. The core concept of this law is <u>mass has the capability of doing work</u>. The core principle of this law is that the amount of energy in the mass of an object is measured and determined by the amount of work done by the object's mass. The core point of this law is: when a mass does positive work, the mass thus decreases; but when a mass does negative work, the mass thus increases. Moreover, the law of mass doing work has revealed the mechanism of/behind the famous and great mass-energy equivalence equation ($E = mc^2$). This mechanism is, and turns out to be that the mc^2 (in $E = mc^2$) is, and is equal to the maximum capability of m doing work; that is to say, this mechanism shows that the mc^2 (in $E = mc^2$) is not only the total energy contained in a rest mass m, but also turns out to be the maximum capability of the rest mass m doing work. And this mechanism further shows that the very reason *why* the total energy contained in a rest mass m is equal to mc^2 is because the maximum capability of the rest mass m doing work is equal to mc^2.

The mass consumption, being a direct product of the law of mass doing work, is the first part of the core point of this law, which is: when a mass does positive work, the mass thus decreases. That is to say, the mass consumption is the <u>decrease</u> in the mass of an object, because the object's mass is consumed (becoming less) due to its doing positive work when the velocity of the object is increased.

The newly developed and verified *mechanism-revealed scales relativity theory*, being the product of the mass consumption (thus being a result of the law of mass doing work), is the *only* scientific theory that has unveiled *why* time runs slower and *why* length becomes shorter at high speed by revealing the <u>mechanism</u> behind these two *whys* or by revealing *why* and *how* the scales of time and length are reduced at high speed.

The mechanism or reason *why* the speed of light is constant has been revealed by the newly developed and verified *mechanism-revealed scales relativity theory* (MRSRT, for short). What should be noticed or realized is that MRSRT is the *only* scientific theory that is capable of answering the question of *why* the speed of light is constant, because MRSRT is the *only* scientific theory that is able to reveal and show *why* time runs slower and *why* length becomes shorter at high speed.

After the mechanism or reason *why* the speed of light is constant has been brought to light, the grave consequences of having been unable to know *why* the speed of light is constant in the past have been analyzed, revealed and shown in the last chapter of this book. These grave consequences are (or can be regarded as) the responses to the possible/potential reactions or opinions, like "Yes, we didn't and couldn't know why the speed of light is constant before, so what?"

All in all, the take-home message about this book is: this book terminates the era in which we human beings haven't known and couldn't know *why* the speed of light is constant. In other words, from now on, the secret of *why* the speed of light is constant will no longer be a secret at all to those who want to know it.

Bingcheng Zhao

Chapter 1

The Questions Unavoidably Pointing to the Law of Mass Doing Work

(The highlight or climax of this chapter)

From the famous and great $E = mc^2$, one can sense or perceive at least three fundamental questions that collectively and consistently, also explicitly and inevitably, point to the law of mass doing work. Thus, the law of mass doing work is not only intrinsically connected to but also essentially based on the known or conventional knowledge; such a feature is very helpful for one to understand this law, though it is unconventionally new in concept; such a feature is also a good indication that it cannot be difficult to understand this law, though it is radically new in concept.

There are at least three fundamentally important questions, which come from $E = mc^2$, that unavoidably and inevitably, also explicitly and consistently, point to the law of mass doing work. What are these fundamentally important questions then? Why do they collectively and consistently, also explicitly and unavoidably, point to this law? Let us analyze, investigate and discuss them in this chapter, being a preparation to introduce this law in the next chapter.

Question one. The famous mass-energy equivalence equation (which is $E = mc^2$, where c is the speed of light, m is the rest mass of an object, and E is the rest energy of the object—being the energy contained in its rest mass m) tells us that mass and energy are equivalent. Since energy can do work—being a universally accepted *fact*, why can't mass? More specifically thus to be more perceptible or noticeable, since energy has the capability of doing work—being a fully recognized *fact*, and since mass and energy are equivalent, it is thus clear, even obvious, that mass also has the capability of doing work; it is necessary and inevitable that mass has the capability of doing work. Therefore, the famous mass-energy equivalence equation (via its mass-energy equivalence and via the universally known *fact* that energy has the capability of doing work) clearly and definitely points to the law of mass doing work (because only this law can reveal, express and reflect the capability of mass doing work).

(*Commentator A: the above analysis is very sharp and penetrating! After thinking it over, I have been completely convinced that the mass-energy equivalence equation clearly points to the law of mass doing work. Moreover, this analysis is not only definitely rational and objective, but also quite simple and clear, thus pretty easy to understand or follow. And so, the above analysis could make one perceive and realize that the famous mass-energy equivalence equation has been looking forward to the law of mass doing work.)

Question two: *why* does mass have energy? The famous mass-energy equivalence equation clearly and explicitly tells us such a greatly important concept: mass has energy; moreover, this concept has been a fundamentally important, fully recognized *fact* in physics or in science for quite a long time. What does such a fundamentally important, fully recognized *fact* (which is mass has energy) clearly and explicitly tell us then? It clearly and explicitly, also unavoidably and undeniably, tells us that the question of *why* mass has energy is definitely and obviously a fundamentally important question in physics or in science. (Commentator B: yes, that's certainly true; that's clearly and surely

2

true! The fundamentally important fact that <u>mass has energy</u> has no choice but to tell us: the question of *why* <u>mass has energy</u> is indeed a fundamentally important question. Moreover, no rational people in the world would deny that the question of *why* <u>mass has energy</u> is a fundamentally important question, as long as they are or have been familiar with the fundamentally important fact that <u>mass has energy</u>.)

To such a fundamentally important question, *why* does mass have energy? The answer from the law of mass doing work is: because mass has the capability of doing work. In fact, only this law can answer such a fundamentally important question, because only this law can reveal, express and reflect the capability of mass doing work. Therefore, the fundamentally important question of *why* <u>mass has energy</u> explicitly and unmistakably points to the law of mass doing work.

(*Commentator B: the above analysis is obviously rational and objective, also quite straightforward, thus very convincing! After deeply pondering the analysis above, I have clearly realized that the question of *why* mass has energy is really and surely a fundamentally important question in physics or in science; I have also sensed that the answer to this fundamentally important question has to depend on the law of mass doing work. As a result, the above analysis indicates or can be understood as that the fundamentally important question of *why* <u>mass has energy</u> is eagerly longing for the law of mass doing work.)

Question three. One can think over the law of mass doing work conversely if necessary: if mass could not have the capability to do work, then it would be scientifically groundless to say that <u>mass has energy</u>! Or what would be the scientific basis to say that <u>mass has energy</u> if mass could not have the capability to do work? (Commentator C: yes, that's undoubtedly correct! If mass could not have the capability to do work, then the well-known fact that mass has energy would actually become totally groundless or baseless in truth.) And so, the well-known *fact* that <u>mass has energy</u> has no choice but to tell us such a clear *fact:* mass has the capability of doing work. As a result—as an explicit and noticeable result in essence, also as an undeniable or irrefutable result in truth, the clear *fact* that mass has the capability of doing work unavoidably and inevitably points to the law of mass doing work (because only this law can reveal, express and reflect the capability of mass doing work).

Moreover, because of the fully recognized and universally known *fact* that energy has the capability of doing work, if mass could not have the capability of doing work, then the fully acknowledged, widely accepted concept of 'mass-energy equivalence' would actually become empty; the so-called 'equivalence' in this concept would inevitably lose its most fundamental, most essential and most important implication; this concept would literally become utterly meaningless. (Commentator C: yes, that's definitely true; that's clearly true!) Therefore, the fully acknowledged, widely accepted concept of 'mass-energy equivalence' also has no choice but to tell us such a clear *fact:* mass has the capability of doing work; this clear *fact* unavoidably and inevitably points to the law of mass doing work (because only this law can reveal, express and reflect the capability of mass doing work).

(*Commentator D: yes, that's clearly and surely true! Either the well-known fact that mass has energy or the fully recognized, widely accepted concept of 'mass-energy equivalence' has no choice but to tell us such a clear fact: mass has the capability of doing work. What's this clear fact doing now? It is eagerly looking for, anxiously waiting for, and earnestly anticipating the law of mass doing work, because only this law can reveal and show the capability of mass doing work.)

All in all, the above three questions collectively and consistently, also explicitly and undeniably, point to the objective and real existence of the law of mass doing work (because these three questions clearly and unavoidably point to the core concept of this law, which is that mass has the capability of doing work). And so, if this law has been discovered, such a discovery ought to be readily accepted. (Commentator E: yes, that's clearly and definitely true! In fact, what this chapter has accomplished is actually calling for the law of mass doing work; this law is also ready to come out once being called.)

Chapter 2

The Law of Mass Doing Work—about Why Mass Has Energy

(The highlight or climax of this chapter)

The (newly discovered and verified) law of mass doing work is the *only* scientific law that answers the fundamental question of **why mass has energy** by revealing and showing that <u>mass has the capability of doing work</u>. The core concept of this law is: <u>mass has the capability of doing work</u>. The core principle of this law is: the amount of energy in the mass of an object is measured and determined by the amount of work done by the object's mass. The core point of this law is: when a mass does positive work, the mass thus decreases; but when a mass does negative work, the mass thus increases.

After the preparation in the last chapter, the law of mass doing work is ready to come out once being called; the task of this chapter is to introduce this law. The law of mass doing work came to the world relatively recently, because it was discovered and verified not very long ago by me, the author of this book.

(*Commentator A: it doesn't matter who discovered the law of mass doing work; in fact, I don't care about the issue of who discovered this law. This is because there is such a generally acknowledged and totally accepted basic principle in science, which is also an objective and rational criterion: for any theory, the things that really matter lie in *what* rather than who—lie with *what* the theory talks about, instead of who developed it. What should be mentioned is that this basic principle has been completely recognized and admitted by the scientific community as general knowledge or common sense in science nowadays; therefore, it is definitely reasonable and realistic to believe that all today's scientists know this basic principle pretty well. On the contrary, if this basic principle were thrown away, science would inevitably lose a rational, objective and fair criterion; the most fundamental nature and spirit of science would be fatally damaged; science would definitely be misled onto a dangerous track; science would no longer be science at all! Please also be reminded or please notice: when Albert Einstein found the famous and great mass-energy equivalence equation, $E = mc^2$, he was neither an important nor influential person in science, actually he was an obscure and insignificant figure in the field of science.) (Commentator B: yes! What commentator A has said above is not only definitely true, but also extremely important to the development and advancement of science, especially to the great or revolutionary breakthroughs in science.)

The core principle of the law of mass doing work is: the amount of energy in the mass of an object is measured and determined by the amount of work done by the object's mass. (Narrator or reminder: such a core principle, when viewed from the angle of comprehension, is quite comparable or very similar to that the energy of a body is measured and determined by the body's capability of doing work in classical physics. Quite obviously, also rather rationally, such a noticeable comparability or similarity is certainly a substantial help for one to perceive and grasp this core principle easily and quickly, which can make one realize or notice that it cannot be difficult to understand the law of mass doing work, even though this law is radically new in concept.) (Commentator C: yes; what the above narrator or reminder has said is rather rational thus quite reasonable. Such a core principle indeed can readily, even easily, enable one to realize or sense: it's not difficult to understand this core principle; accordingly, it's not difficult to understand the law of mass doing work, though this law is quite new in concept.)

Concisely, as the exact reflection of this core principle, the law of mass doing work (with accurate mathematical expression) reveals and shows: when the velocity of an object is increased, the object's mass does positive work, the object thus loses the same amount of energy as that of the work done by the mass of the object from and by consuming its mass. As a result, the core point of this law is: an object's mass doing positive work causes a corresponding decrease in the object's mass; that is, when a mass does positive work, the mass thus decreases. (One can clearly and easily understand this core point via such a simple comparison in classical physics: when a body does positive work, the energy of the body decreases.) (Commentator D: such a core point is very helpful for one to perceive and realize: it cannot be difficult to comprehend this core point; it thus cannot be difficult to comprehend the law of mass doing work, though it is a newly discovered physical law.) The other side of this core point is: an object's mass doing negative work, which occurs when the velocity of the object is decreased, causes a corresponding increase in the object's mass; that is, when a mass does negative work, the mass thus increases. (One can clearly and easily comprehend this aspect via such a simple comparison in classical physics: when a body does negative work, the energy of the body increases.) (Commentator E: moreover, if one thinks of the core principle and the core point of the law of mass doing work simultaneously, it seems quite reasonable to believe that he or she will have no difficulty realizing it is actually pretty easy to understand this law, albeit it is a newly developed physical law.)

*Friendly reminder: dear readers, if you are the professional people in physics, especially in modern physics, you can comprehend the law of mass doing work more easily and quickly than others. This is because the law of mass doing work directly and totally comes from the relativistic kinetic energy of an object with rest mass m (by exchanging the positions of the velocity v and the relativistic momentum p in the integral calculation of the relativistic kinetic energy of the object, then by the definite integral operation from 0 to v with velocity v as variable). As a result, the relativistic kinetic energy (of the object) is the area that is *under* the line of a certain relativistic momentum p (at a certain velocity v) and *above* the line of the relativistic momentum p (i.e., the area between these

p

p

A1

A2

v/c

two lines, which is A1.   where c is the speed of light); whereas the law of mass doing work is the area that is *under* the line of the relativistic momentum p, which is A2. That is to say, the law of mass doing work is not only connected to but also based on the known or conventional knowledge; such a feature can substantially enhance the recognition and acceptance of this law, though it is unconventionally new. Moreover, the law of mass doing work can be

8

simply expressed as the product of the force acting on an object with rest mass m and the displacement of the object (note: the product of quantity A and quantity B is A times B); i.e., the expression of this law is completely consistent with the fully recognized expression of the work done by a force in classical physics. Quite obviously, also rather rationally, such a complete consistency can not only substantially but also explicitly enhance the recognition and acceptance of this law, even though it is a newly discovered physical law.

What should be pointed out is that the concept and equation of the relativistic kinetic energy of an object with rest mass m have been written into the textbooks for college or graduate education (since far more than half a century ago); that is, this concept and equation have already been fully recognized and accepted by the scientific community in physics. And so, the professional people in physics or in modern physics, because they have been familiar with this concept and equation very well, really have an obvious advantage over others in compre-hending the law of mass doing work, which can make them compre-hend this law much more easily and quickly than others, even though the discovery of this law might be radically new in the eyes of some of those respected conventional professional people. So my sincere congratulations go to the professional people in physics for this obvious advantage; please accept my sincere and rational congratulations if you are the professional people. (An independent and rational reviewer: because the law of mass doing work directly and entirely comes from the equation of the relativistic kinetic energy of an object with rest mass m; because this equation has been fully recognized and completely accepted by the scientific community in physics; because nothing is added or removed in the process from this equation to the law of mass doing work, there is neither rational reason nor valid basis not to accept the law of mass doing work. In fact, there is utterly no way to deny the law of mass doing work from the perspective of science.)

Chapter 3

Revealing the Mechanism of/behind the Mass-Energy Equivalence Equation

(The highlight or climax of this chapter)

As a direct and greatly important application of the law of mass doing work, this law has revealed the mechanism of/behind the mass-energy equivalence equation ($E = mc^2$). This mechanism is, and turns out to be that the mc^2 (in $E = mc^2$) is, and is equal to the maximum capability of m doing work; that is, this mechanism shows that the mc^2 (in $E = mc^2$) is not only the total energy contained in a rest mass m, but also turns out to be the maximum capability of the rest mass m doing work. And this mechanism further shows that the very reason *why* the total energy contained in a rest mass m is equal to mc^2 is because the maximum capability of the rest mass m doing work is equal to mc^2.

B eing a direct, also tremendously important, application of the law of mass doing work introduced in the last chapter, this law has revealed the mechanism of/behind the famous mass-energy equivalence equation (this equation is $E = mc^2$, where c is the speed of light, m is the rest mass of an object, and E is the rest energy of the object—being the energy contained in its rest mass m. This famous equation is often referred to as the greatest equation in the history of science). Then what does this mechanism turn out to be after revealing it?

This mechanism turns out to be: the rest energy of an object, being the total energy contained in the rest mass of the object, is equal to the maximum capability of the object's rest mass doing positive work (this maximum capability is equal to mc^2, being the right side of the famous $E = mc^2$). So this mechanism shows that the world-famous mc^2 (in the famous $E = mc^2$) is not only the total energy contained in a rest mass m, but also turns out to be the maximum capability of the rest mass m doing (positive) work. Moreover, this mechanism further shows that the very reason *why* the total energy contained in a rest mass m is equal to mc^2 is because the maximum capability of the rest mass m doing (positive) work is equal to mc^2.

(*Related question and answer: what is the profound and essential difference <u>before</u> and <u>after</u> revealing the mechanism of/behind the famous and great mass-energy equivalence equation? Answer: <u>before</u> this revealing, we human beings merely knew mass has energy, but couldn't know *why;* <u>after</u> this revealing, we human beings know *why* mass has energy via knowing that mass has the capability of doing work. Accordingly, also unavoidably, this profound and essential difference is also an explicit demonstration or clear reflection of the fundamental importance of the law of mass doing work, corresponding to the fundamentally important status of this famous and great equation in science.)

After revealing the mechanism of/behind the famous and great mass-energy equivalence equation, the solid existence of this mechanism is an irrefutable fact, a bit like: having found out the continent of North America is the irrefutable fact that there is this continent; this is also a bit like: having found out diamond beneath a certain place is the hard evidence that there is diamond beneath this place. (Commentator A: yes, no rational people in the world want to deny the solid existence of this mechanism, simply because it has been revealed; of course, no one can deny the solid existence of this mechanism, because and <u>after</u> this mechanism has been revealed.) Not only that, <u>after</u> revealing the mechanism of/behind this famous and great equation, its validity and reliability thus become further solid and secure—because it turns out that this famous and great equation does have a very solid and secure mechanism. (Narrator: and so, from now on—from the moment when its mechanism has been revealed, the famous and great $E = mc^2$ could confidently and bravely declare to the entire world—with sufficient and irrefutable evidence: I do have a solid and secure mechanism!)

(Commentator B: wow! Aha! The mechanism of/behind the famous and great mass-energy equivalence equation is finally brought to light. This is definitely a remarkably important event to this great equation, because this mechanism, only this mechanism, is able to answer the biggest *why* underlying this great equation, *why* mass has energy—because mass has the capability of doing work! Moreover, if one thinks over this mechanism for a few minutes, it seems not difficult that he or she could clearly and surely realize: only the law of mass doing work is able to reveal the mechanism of/behind this great equation, believe it or not. This realization can readily, even easily, enable one to be aware that the existence of this law turns out to be an explicit *fact*, a clear *fact*, also an undeniable or irrefutable *truth*, because the existence of this famous and great equation has become a universally acknowledged, well-known *fact;* because only this law can reveal the mechanism of/behind this very equation. Thus, even if Bingcheng Zhao, the author of this book, had not discovered this law, somebody else would find it someday, sooner or later; the earlier, the better, of course. Yet regardless of who has discovered this law, the famous mass-energy equivalence equation is or should be equally happy, because this law has revealed the mechanism of/behind this famous equation, which is also often referred to as the greatest equation in the history of science. After this revealing, this famous and greatest equation, via unfolding and displaying its great mechanism that answers the biggest *why* underlying this greatest equation, also the most fundamental *why* underlying this greatest equation, appears and becomes even more beautiful and charming.)

When the famous mass-energy equivalence equation, or the greatest equation in the history of science, is at the age of more than one hundred years old, its secret veil is finally unveiled—the mechanism of/behind this famous and great equation has been at last revealed. (Commentator C: fortunately, this famous and great equation is not a bride! Of course, if it had been a bride, probably no bridegroom in the world would have been patient enough to wait for such a long time to unveil her veil after their wedding. But for a fundamentally important and extraordinarily influential equation in science like this famous and great equation, it seems that the later unveiling of its secret veil, the more wonderful its marvelous charm is, being a bit like wine: a bottle of old wine is more tasteful and mellower than a new one.) In this sense, the famous and great mass-energy equivalence equation should be happy for itself—be happy for its great mechanism having been finally revealed. In this sense, this famous and great equation ought to congratulate itself—congratulate its great mechanism having been at last revealed. In this sense, this famous and great equation must celebrate itself—celebrate its great mechanism having been finally brought to light! Moreover, and in a broad sense, it seems acceptable if this famous and great equation wants to invite all the people in the world, especially those respected and related experts in physics, to have a grand and solemn celebration of this great and historic revealing! (Most probably, this famous and great equation will provide delicious food and excellent wine for all of us in such a spectacular, splendid, and wonderful occasion.)

More than what we have seen above, the very fact, which is that the law of mass doing work has revealed the mechanism of/behind the famous mass-energy equivalence equation, clearly and definitely points to the great importance of this law along the following explicit and noticeable direction. Since this famous equation is widely recognized as the greatest equation in science, then its mechanism is, or ought to be, the greatest mechanism in science; since the law of mass doing work is the only physical law that reveals this greatest mechanism, then it is actually rational and appropriate (or at least it is neither irrational nor inappropriate) if one comes to the conclusion that this law is the greatest law or one of the most fundamental and most important laws in science. (Commentator D: yes; this conclusion is obviously rational and objective, thus undoubtedly appropriate. And so, it is no exaggeration to say that the discovery of the law of mass doing work is really and truly a great achievement or historic event in science, believe it or not.) (Correspondingly, what readers have seen above is, or ought to be, the greatest law or one of the most fundamental and most important laws in science; so the author of this book genuinely congratulates you, dear readers.)

(*Related question and answer: because the law of mass doing work has revealed the mechanism of/behind the famous and great mass-energy equivalence equation, can this very equation itself be a good window, through which one could see this law more clearly, thus have a better understanding of this law? Or can this famous and great equation substantially help one understand this law impressively and explicitly? Answer: yes, it can; please see the specific and closely related facts or information in the coming paragraph.)

One could tangibly and quickly grasp the law of mass doing work if he or she views this law from the following several important, also easily perceptible, angles. Angle A, from the large perspective of the fundamental question, *why* does mass have energy? The answer from this law is: because mass has the capability of doing work; in fact, only this law can answer such a fundamental question. Angle B, the famous mass-energy equivalence equation tells us that mass and energy are equivalent; and since energy can do work—being a universally accepted *fact*, why can't mass? More specifically, since energy has the capability of doing work—being a fully recognized *fact*, and since mass and energy are equivalent, it is clear, even obvious, that mass also has the capability of doing work; it is necessary and inevitable that mass has the capability of doing work. (Otherwise, the fully acknowledged concept of 'mass-energy equivalence' would lose its most fundamental, most essential and most important implication; this concept would thus become meaningless in fact.) Angle C, one can think over this law conversely if necessary: if mass could not have the capability to do work, it would be scientifically groundless to say that mass has energy! In other words, the known *fact* that mass has energy has no choice but to tell us another inevitable *fact:* mass has the capability of doing work. Angle D, since mass has the capability of doing work, it becomes quite natural that, when an object's mass does *positive* work, the object's mass thereby *decreases* (one can clearly perceive and easily comprehend this point if he or she is familiar with such common knowledge in classical physics: when a body does *positive* work, the available energy of the body thus *decreases*). (Commentator E: when one views the law of mass doing work through the diverse visual angles like the above, he or she could see this law more clearly from different directions, a bit like 3-D visual effects; he/she could thus grasp this law more tangibly and effectively. Once one has grasped this law, he or she could also clearly and easily realize that this law can lay the solid foundation for any theories based on it.)

Last but not least, what should be mentioned is that the above-mentioned fact, which is that the law of mass doing work has revealed the mechanism of/behind the famous mass-energy equivalence equation, is also fundamentally and crucially important to this newly discovered law. This fundamental and crucial importance is explicitly reflected in such a clear *fact:* this law has been verified or confirmed via its revealing the mechanism of/behind the famous mass-energy equivalence equation, because this famous equation has passed experimental tests many, many times since its birth, thus being a fully recognized and universally accepted *fact*. With this verification or confirmation, the validity and reliability of this law are thus quite positive. (So, for this verification or confirmation, this law, on behalf of the discoverer of this law, wants to express its sincere acknowledgment to all the related scientists for their great contributions that have made this famous and great equation found and verified, especially to the great scientist Albert Einstein, the founder of this equation.)

Chapter 4

The Mass Consumption

(The highlight or climax of this chapter)

The mass consumption, as revealed by the law of mass doing work, is the *decrease* in the mass of an object, because the object's mass is *consumed* (becoming less) due to its doing *positive* work when the velocity of the object is increased; that is, the mass consumption occurs (to the mass of an object) when the velocity of the object is increased. Besides, as long as one has known *why* mass has energy or *why* $E = mc^2$, he or she can and will clearly and easily understand the mass consumption.

The mass consumption (being the direct result of the combination of the two things introduced in the last two chapters: the law of mass doing work and the mechanism of/behind the famous mass-energy equivalence equation revealed with this law) shows that the mass of an object *decreases* with the increase in its velocity, by revealing *why* and *how* the object's mass is being consumed due to its doing positive work in such a situation. Concisely, the mass

17

consumption is the *decrease* in the mass of an object, because the object's mass is consumed (becoming less) due to its doing positive work when the velocity of the object is increased. Therefore, the mass consumption, being caused by mass doing positive work, is simply the product of an application of the newly discovered and verified law of mass doing work. (Related question and answer: how could one understand the mass consumption clearly and easily, also effectively and impressively? Answer: please view and think over the mass consumption from the following three important aspects or angles.)

The first aspect or angle: from the fact that mass has the capability of doing work. As analyzed, revealed and shown in the earlier chapters, mass has the capability of doing work. Since mass has the capability of doing work, it is rather natural and quite reasonable that, when an object's mass does *positive* work, the object's mass thereby *decreases* (one can clearly perceive and easily comprehend this aspect via the considerable and explicit help from such a comparable concept in classical physics: when a body does *positive* work, the available energy of the body thus *decreases*).

The second aspect or angle: from the angle of the mass-energy equivalence equation. As long as one has known or heard of the famous mass-energy equivalence equation, which is $E = mc^2$ (where c is the speed of light, m is the rest mass of an object, and E is the rest energy of the object—being the energy contained in its rest mass m), he or she could clearly and impressively comprehend the mass consumption, because it is inherently connected to the mechanism of/behind this famous equation. To be further explicit, both the mass consumption and this famous equation are attached onto the same thing—the law of mass doing work, because the mass consumption, being caused by *mass doing positive work*, directly and totally comes from this law; because the mechanism of/behind this famous equation, as revealed by this law and as presented in chapter three, is the maximum capability of a rest mass m *doing positive work*. That is, both the mass consumption and this famous equation have the same mechanism—*mass doing positive work*. As a result, this great and famous equation turns out to be a great and explicit help for one to understand the mass consumption clearly and easily, also impressively.

The third aspect or angle: from the angle of the known concept and equation. The fully recognized and completely accepted concept and equation of the relativistic kinetic energy of an object with rest mass m can also provide a substantial and explicit help for one to understand the mass consumption. This is because the law of mass doing work directly and totally comes from this concept and equation, as clearly pointed out in chapter two; and this is because the mass consumption is simply the product of an application of this law, as explicitly mentioned in the first paragraph of this chapter.

All in all, one could understand the mass consumption clearly and easily, also effectively and impressively, if he or she views and thinks over the mass consumption from the three important aspects or angles above. That is to say, with and through the substantial and explicit help from these aspects or angles, it is or can be rather rational or quite reasonable to conclude or believe that one will have no difficulty perceiving and understanding the mass consumption (or at least will have no difficulty clearly realizing the objective and real existence of the mass consumption).

Even having understood the mass consumption clearly and impressively (through what have been seen from the three important aspects or angles above), some careful readers, especially some dear readers with rich knowledge in modern physics, might think of the concept of relativistic mass (the so-called relativistic mass is an important concept formed within the paradigm of special relativity. This concept says that the mass of an object increases with the increase in its velocity, and the mass of an object becomes infinitely large when the object infinitely approaches the speed of light; that is, mass increases with speed. So 'relativistic mass' is often simply said as 'rest mass is least' in the various materials on special relativity); and these readers might have noticed or realized that the mass consumption and relativistic mass are opposite. And so, it seems better that I should provide a relevant clarification here for avoiding possible confusion. This clarification is: there are fundamental and obvious differences between the mass consumption and relativistic mass.

Concisely, these differences are specifically reflected in the following three fundamentally important aspects. (i) The mass consumption is inherently connected with the mechanism of/behind the famous and great mass-energy equivalence equation ($E = mc^2$), because the theoretical basis of the mass consumption, which is the law of mass doing work, has also revealed this mechanism, as introduced in chapter three. This inherent connection is a clear or good indication of the validity and correctness of the mass consumption. On the contrary, relativistic mass

20

has nothing to do with the mechanism of/behind this famous and great equation, because relativistic mass was the product long before the law of mass doing work had been discovered; in fact, relativistic mass literally prevents from revealing the mechanism of/behind this famous and great equation—that is, relativistic mass prevents human beings from knowing *why* mass has energy, because only this mechanism can tell us *why* mass has energy, as emphasized or mentioned in the last three chapters. In other words, the mechanism of/behind the famous and great mass-energy equivalence equation (thus this famous and great equation) has to say NO to relativistic mass, believe it or not; that is to say, anyone, as long as he or she has accepted the famous and great $E = mc^2$ and has known its mechanism, should/could explicitly perceive and realize such a clear *fact:* relativistic mass can neither be valid nor correct in truth. (ii) The mass consumption is totally consistent with such a fundamental principle in classical physics: when a body does *positive* work, the available energy of the body thus *decreases.* In contrast, relativistic mass is literally at odds with this fundamental principle, because according to relativistic mass, when a mass does *positive* work, the mass thus *increases*, which is obviously and flatly ridiculous; that is, relativistic mass is obviously and flatly ridiculous. (iii) The mass consumption is the indispensable theoretical basis of the new theory that reveals and shows *why* time runs slower at high speed (this new theory, which is mechanism-revealed scales relativity theory, will be introduced in the next chapter); whereas relativistic mass turns out to be an absolutely impassable obstacle to developing such a new theory—that is, within the paradigm of relativistic mass, it is definitely impossible for human beings to know the secret of *why* time runs slower at high speed. All in all, with and through these fundamental and obvious differences, it seems rather rational thus quite reasonable to conclude and/or believe that one could get rid of the hindrance or interference from relativistic mass in comprehending the mass consumption.

Chapter 5

The New Theory That Reveals
Why Time Runs Slower at High Speed

(The highlight or climax of this chapter)

The long-standing, fundamental question of *why* time runs slower at high speed has been finally answered, and answered by the newly developed and verified theory introduced in this chapter. Moreover, this new theory is the *only* scientific theory that unveils *why* time runs slower and *why* length becomes shorter at high speed by revealing the mechanism behind these two *whys* or by revealing *why* and *how* the scales of time and length are reduced at high speed.

After seeing the title of this chapter, especially after reading what has appeared in (The highlight or climax of this chapter), some, even many, readers perhaps feel bewildered. They may have the bewilderment like: we have known that time runs slower and length becomes shorter in the situation of high speed from Einstein's theory of special relativity, why are you going to introduce a new theory? Other readers may even have the reaction like: because we have known that time runs slower and length becomes shorter at high speed from Einstein's theory of special relativity, we don't need you to introduce a new theory at all. This kind of possible bewilderment or reaction seems to call for a closely related clarification, which is presented as follows.

This closely related clarification is: while Einstein's theory of special relativity does tell us that time runs slower and length becomes shorter in the situation of high speed, it is really unable to solve the problems of *why* time runs slower and *why* length becomes shorter in such a situation (or unable to answer the questions of *why* time runs slower and *why* length becomes shorter in such a situation), simply because it is incapable of revealing the mechanism behind these two *whys*.

Specifically and evidently, one could clearly perceive and explicitly realize, even only from the basic and prominent feature of special relativity, the hard *fact* that special relativity is really unable to solve the problems of *why* time runs slower and *why* length becomes shorter in the situation of high speed, as long as he or she views or thinks over this hard *fact* in the simple and clear way like the following. If special relativity had been able to solve these two problems, its first postulate (this postulate says that the speed of light is the same for all observers, regardless of their motion relative to the source of light, or being simply referred to as 'the constancy of the speed of light' or as 'the constant speed of light') would not have been necessary at all. And so, the irrefutable or undeniable *actuality* that special relativity <u>indispensably</u> and <u>desperately</u> **necessitates** its first postulate can enable one to perceive this hard *fact* clearly and realize it explicitly; of course, this irrefutable or undeniable *actuality* also unmistakably shows or unavoidably points to: one has no way to deny this hard *fact*, or no one can deny this hard *fact*.

Not only that, if special relativity had been able to solve these two problems, its first postulate would no longer have been a postulate; along with the constant and specific reminder from such a naked *truth:* within the paradigm or stereotype of special relativity, its first postulate is <u>always</u> a postulate! As a result, this naked *truth* can help one perceive and realize this hard *fact* even more clearly and explicitly.

24

Moreover, with this naked *truth*, no one can deny this hard *fact*. (Friendly reminder: after revealing the mechanism of/behind the first postulate of special relativity with a new theory—this revealing will be introduced in the next chapter, dear readers will see, through a sharp contrast, this naked *truth* thus this hard *fact* even more clearly.)

More than that, the first postulate of special relativity, or its postulate of 'the constancy of the speed of light', being one of the most famous, most important and most influential postulates in modern physics, turns out to be actually an impartial eyewitness and unforgettable reminder of the definite existence of this hard *fact*, which can make or help one face this hard *fact* bravely and rationally, perceive it clearly and impressively, realize it explicitly and confidently.

More concisely, one could also clearly perceive, explicitly realize and unmistakably understand the hard *fact* that special relativity is indeed unable to solve the problems of *why* time runs slower and *why* length becomes shorter at high speed via the following straightforward and simple method. Because special relativity is unable to answer the question of *why* the speed of light is constant, it doesn't have the ability to answer the questions of *why* time runs slower and *why* length becomes shorter at high speed (being the same thing as that, because special relativity is unable to solve the problem of *why* the speed of light is constant, it doesn't have the ability to solve the problems of *why* time runs slower and *why* length becomes shorter at high speed).

Most concisely, the theory of special relativity, because it is unable to know *why* the speed of light is constant, doesn't have the ability to know *why* time runs slower and *why* length becomes shorter at high speed.

*Please accept my sincere congratulations, if you are the professional people in physics, especially in modern physics, because you have an obvious advantage over others in comprehending the clarification above. The obvious advantage is that you are very familiar with the first postulate of special relativity, or its postulate of 'the constancy of the speed of light'. This obvious advantage can enable you to comprehend the above clarification much more easily and quickly than others.

Even with the clarification above, some readers may still be baffled. These readers may think of or ask a related question like: it is often said that special relativity has passed observational tests, *why* is it still unable to solve the problems of *why* time runs slower and *why* length becomes shorter at high speed? This kind of bafflement or question seems to call for another related clarification.

This clarification is to be done through knowing or reviewing the pertinent, also greatly important, general knowledge or common sense in science, which is that observational or experimental tests themselves have neither the function nor the ability to answer the questions about *whys* or solve the problems about *whys*. In fact, the task or purpose of observational or experimental tests is not to deal with these questions or these problems at all; instead, the task of answering the questions about *whys* or solving the problems about *whys* is, or is supposed to be, responsible by theories. In other words, if a theory is unable to solve the problems about *whys*, please don't expect or think that observational or experimental tests of the theory can solve these problems. Undoubtedly, such general knowledge or common sense can make or help one see even more clearly and explicitly the hard *fact* that special relativity is unable to solve the problems of *why* time runs slower and *why* length becomes shorter at high speed. (Reviewer: even having presented the two clarifications above, please don't forget to mention such a reality: the problem or question of *why* time runs slower in the situation of high speed has been perplexing many brilliant and curious physicists since the birth of special relativity in 1905. So this reality actually tells us: it has already become a clearly perceived or realized *fact* that special relativity is unable to answer the question of *why* time runs slower at high speed. As a result, the real implication of this reality is either closely related to or consistent with these two clarifications, which can greatly help one face and understand them.)

Even with the two clarifications above, the hard *fact*, which is that special relativity is unable to solve the problems of *why* time runs slower and *why* length becomes shorter at high speed, may still be a surprise to some dear readers. For reducing such a possible surprise, it seems not inappropriate that I shall provide a clear clue to help these readers face or digest this hard *fact*. This clear clue directly comes from the explicit, also noticeable, feature of special relativity.

This feature is that the starting point of special relativity is its first postulate, simply referred to as 'the constancy of the speed of light' (this postulate has just been mentioned above). Please carefully notice that the speed of light is a *derived*, composite quantity, rather than a *fundamental* quantity—the speed of light is the distance it has traveled divided by the time it has taken. And more candidly thus more plainly, the starting point of special relativity is merely the tactic of "forcing" two *fundamental* quantities (time and length) to match up a *derived*, composite quantity (the speed of light). (Commentator: yes, that's true. In fact, one can safely say that all the professionals in physics are well aware of such a basic knowledge: time and length are two fundamental quantities, whereas speed or velocity, no matter whether it is the speed of a car or a plane or even a projectile, of course including the speed of light, is a derived, composite quantity.) As a result, this noticeable feature can easily enable us to think of or think over: given that the starting point of special relativity is a derived, composite quantity, rather than a fundamental one; whereas time and length (both are the core quantities that special relativity deals with) are fundamental quantities, and so, it is readily understandable, at least not a big surprise at all, that special relativity is unable to solve the problems that are exactly about time and length. All in all, this explicit and noticeable feature can considerably and impressively help one perceive and realize the hard fact that special relativity is unable to solve the problems of *why* time runs slower and *why* length becomes shorter at high speed.

What does this hard fact really tell us? Let us see it. Since the major subject of special relativity is to tell us that time runs slower and length becomes shorter in the situation of high speed (this subject has many other similar expressions or descriptions, like: both time and length change with speed; time and length are variable thus relative at different speeds; both time and length are different at different speeds, etc.), clearly, even obviously, the problems of *why* time runs slower and *why* length becomes shorter in such a situation are definitely the fundamentally important problems in front of special relativity. In other words, special relativity turns out to be a theory that is unable to solve the fundamentally important problems in front of it, because it is incapable of revealing the mechanism behind these two *whys*.

All in all, the pieces of information presented above (including two clarifications and a closely related clear clue) clearly point to the hard fact that special relativity is really unable to solve the fundamentally important problems: *why* time runs slower and *why* length becomes shorter in the situation of high speed. With this hard fact kept in mind, it seems not difficult that one can realize that it's definitely necessary or obviously important to introduce the new theory that unveils *why* time runs slower and *why* length becomes shorter at high speed by revealing the mechanism behind these two *whys*. Let us go to this new theory. (Narrator: this new theory is pleasantly and enthusiastically waiting for us; a warm welcome is awaiting all of us! This new theory is always very hospitable to all visitors and readers.)

After the related clarifications above, we are ready to go to this new theory, which is the result of an application of the mass consumption caused by mass doing positive work (the mass consumption has been introduced in the last chapter). Since a mass meter (the equipment used to measure mass, a balance, for example) consists of different blocks with standard mass, the mass consumption is, of course, applicable to mass scale—the mass of all standard blocks is consumed or reduced according to the mass consumption, referred to as mass scale reduction. Because any unit length of a meter stick (that is, the length scale of a meter stick, such as per centimeter on a meter stick) is made of a certain amount of mass, clearly the mass consumption is equally applicable to length scale too—the mass between two neighboring graduations on a meter stick is consumed or reduced according to the mass consumption, referred to as length scale reduction. Because any unit time of a clock (that is, the time scale of a clock such as per minute) is composed of a certain amount of mass, clearly the mass consumption is also equally applicable to time scale—the mass between two neighboring graduations on a clock is consumed or reduced according to the mass consumption, referred to as time scale reduction. As a result, due to the effect of the mass consumption caused by mass doing positive work, the scales of mass, length and time in the situation of moving at a certain high speed are <u>all</u> reduced <u>at the same rate</u> with respect to the scales in the situation of not moving or moving at a lower speed (Fig. 5.1). (The situation of not moving refers to that there is always no position change with respect to a third independent, stationary reference point.)

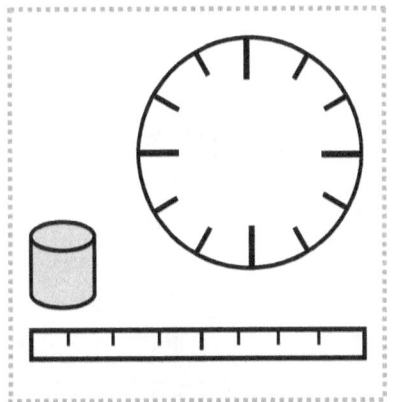

The scales of mass, length and time in the situation of *not moving or moving at a much lower speed;* that is, the mass scale, length scale and time scale in such a situation.

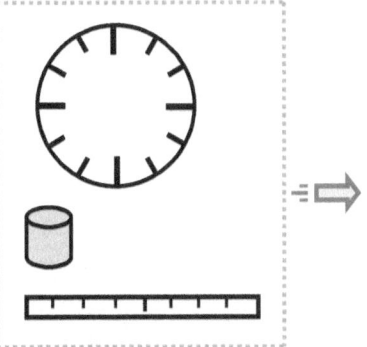

The scales of mass, length and time in the situation of *moving at a certain very high speed;* that is, the mass scale, length scale and time scale in such a situation.

Figure 5.1, why time runs slower in the situation of high speed (or why and how the scale of time is reduced in such a situation); why length becomes shorter in the situation of high speed (or why and how the scale of length is reduced in such a situation).

Specifically, also concisely, this new theory, through its time scale reduction and length scale reduction, accomplishes the unprecedented task with profound implications and historic significance: unveiling *why* time runs slower and *why* length becomes shorter in the situation of high speed (still Fig. 5.1). For instance, time scale reduction unveils the reason why a moving clock runs slower than a clock at rest is: a stationary observer reads the moving, scale-reduced clock with his own larger scale. This is a bit like that, when you read a clock of 8-centimeter face according to the time scale on a clock of 10-centimeter face, you will find that the clock of 8-centimeter face runs slower (than the clock of 10-centimeter face; you can draw two such clocks on two transparent papers, and read them). Therefore, time scale reduction unveils *why* time runs slower at high speed by revealing the mechanism behind this *why*. Similarly, length scale reduction unveils the reason why a moving meter stick becomes shorter than a meter stick at rest is: a stationary observer reads the moving, scale-reduced meter stick with his own larger scale. Therefore, length scale reduction unveils *why* length becomes shorter at high speed by revealing the mechanism behind this *why*. (Related reminder: in contrast, special relativity is unable to tell us *why* time runs slower and *why* length becomes shorter at high speed, as analyzed and clarified above. Thus, there is a funda-mental, also crucial, difference between this new theory and special relativity. Such a difference is also a clear demonstration or explicit indication of the obvious necessity and/or great importance to introduce this new theory.)

What we have seen above is the most important basic content of this new theory, also its pivotal or core content. Because it reveals and determines <u>why and how</u> mass, length and time are variable thus relative at different speeds by finding out the relationship in their scales (or because it reveals and shows <u>why and how</u> the scales of mass, length and time in the situation of moving at a certain high speed are <u>all</u> reduced <u>at the same rate</u> with respect to the scales in the situation of not moving or moving at a lower speed); because its theoretical foundation is of mechanism-revealed nature; and because it reveals the mechanism behind its describing phenomena, this new theory has been named as mechanism-revealed scales relativity theory (MRSRT, for short) by me, the founder of MRSRT. The take-home message about MRSRT is: MRSRT unveils *why* time runs slower and *why* length becomes shorter in the situation of high speed by revealing the mechanism behind these two *whys;* that is, MRSRT resolves the fundamentally important problems of *why* time runs slower and *why* length becomes shorter at high speed. (Commentator or reminder: as a result, the birth of MRSRT terminates the era in which we human beings haven't known and couldn't know *why* time runs slower at high speed. From now on, we human beings will no longer be baffled by the long-standing, fundamental question of *why* time runs slower at high speed. Let us together warmly embrace this great, historic era!)

*Related questions and answers: what is the quickest route or one of the quickest routes to go into MRSRT? Or how could one enter MRSRT easily, quickly and naturally from the recognized or available knowledge? Answer: as long as one has thought of the famous mass-energy equivalence equation (again which is $E = mc^2$, where c is the speed of light, m is the rest mass of an object, and E is the rest energy of the object—being the energy contained in its rest mass m), which is often referred to as the greatest equation in the history of science, he or she could enter MRSRT easily, quickly and naturally. This is because the fundamental theoretical foundation of MRSRT—the law of mass doing work has also revealed the mechanism of/behind this famous and greatest equation, as seen in chapter three; thus, essentially speaking, the mechanism of/behind this equation is ultimately this law. That is, the mechanism underlying this equation turns out to be exactly the same thing as the fundamental theoretical foundation of MRSRT. And so, MRSRT and this equation are attached onto the same thing—the law of mass doing work. Therefore, this famous and greatest equation turns out to be an eye-catching sign that directs one to go into MRSRT: once one has seen this sign, he or she will have no difficulty entering

MRSRT. Or stated concisely, the direct theoretical foundation of MRSRT, which is the mass consumption, is based on the fundamental theoretical foundation of MRSRT (which is the law of mass doing work) that has revealed the mechanism of/behind the famous and great $E = mc^2$, and that has shown us *why* mass has energy. Question: what are variable thus relative in MRSRT? Answer: time and length are variable thus relative. Question: why are they variable thus relative? Answer: because their scales are variable thus relative. Question: what is the root cause or the most fundamental reason that time and length are variable thus relative in MRSRT? Answer: the root cause, also the most fundamental reason, is the law of mass doing work, because MRSRT directly comes from the mass consumption that directly and totally comes from this law. In other words, this root cause and the famous mass-energy equivalence equation are attached onto the same thing— the law of mass doing work, because this law has also revealed the mechanism of/behind this famous equation, as just mentioned above. This feature can make or help one grasp this root cause easily, quickly and firmly, because this famous and greatest equation is so well known that many people have known it, or at least have heard of it. In addition, what should be noticed or realized is that space is also variable thus relative in MRSRT, because length scale reduction is equally applicable to width scale reduction and height scale reduction (note: length, width and height are the three dimensions that determine the size of space).

Chapter 6

Unveiling the Long-Standing Secret of Why the Speed of Light Is Constant

(The highlight or climax of this chapter)

The long-standing question of *why* the speed of light is constant has been finally answered by revealing the mechanism behind this *why* (or answered by finding out the reason or cause behind this *why*); that is, the mechanism behind the first postulate of special relativity has been finally revealed. Moreover, the mechanism behind the second postulate of special relativity has also been revealed (this postulate says that all observers moving at constant speed should have the same physical laws). Therefore, from now on, the two postulates of special relativity will no longer be postulates at all, simply because the mechanism behind them has been revealed. What should be mentioned is that only the new theory introduced in the last chapter has the ability to reveal the mechanism behind these two extraordinarily important postulates.

The task of this chapter is to introduce *why* and *how* the mechanism behind the two postulates of special relativity has been revealed (that is, the reason or cause behind these two postulates has been discovered), and revealed by the newly developed and verified theory introduced in the last chapter, which is mechanism-revealed scales relativity theory. (Narrator or commentator: the revealing of the mechanism behind the two postulates of special relativity finally terminates the era in which these two postulates have been postulates for more than one century; that is to say, from now on, these two highly famous and greatly important postulates will no longer be postulates at all, simply because the mechanism behind them has been revealed.)

Let us first see *why* and *how* the mechanism behind the first postulate of special relativity has been revealed, because it is this mechanism that tells us *why* the speed of light is constant (that is, it is this mechanism that answers the long-standing, big question of *why* the speed of light is constant, because this mechanism is the reason or cause behind this *why*).

As mentioned in the last chapter, the first postulate of special relativity says that the speed of light is the same for all observers, regardless of their motion relative to the source of light, or being simply referred to as 'the constancy of the speed of light'. This postulate is either the most important postulate or one of the most important postulates in modern physics, also being one of the most influential and famous postulates in modern physics. This is determined by the fundamentally important status of this postulate in the theory of special relativity as well as the fundamentally important status of special relativity in modern physics. (Related knowledge: special relativity is the crucial theoretical basis of general relativity. General relativity and quantum mechanics are the two main theoretical pillars of modern physics. Or sometimes special relativity, general relativity and quantum mechanics are simply referred to as the three main theoretical pillars of modern physics.)

Specifically, of the two postulates of special relativity (its second postulate will be seen soon in this chapter), while either of them is indispensable to special relativity, its first postulate is more important, thus being often referred to as the leading postulate (Einstein himself made or called it a key postulate. Please be reminded: when developing special relativity, Einstein first accepted 'the constancy of the speed of light' as a fact—that is, he accepted its first postulate as a fact, and then tried every method or scheme to fit this fact). Moreover, according to Max Planck (the earliest founder of quantum theory; he suggested the quantum hypothesis that initiated the quantum idea in 1900, the year in which the era of modern physics began), the velocity of light, or the first postulate of special relativity, is more or less equivalent to the theory of special relativity. All in all, the first postulate of special relativity is fundamentally and crucially important to special relativity—without this postulate, there would have been no special relativity at all.

In the face of such a fundamentally and crucially important postulate, it seems neither unusual nor irrational if one thinks of or asks funda-mental, also crucial, questions like: is there and/or what is the mechanism behind this postulate? Clearly, if there is the mechanism behind this postulate, this mechanism ought to be fundamentally and crucially important, corresponding to the fundamentally and crucially important status of this postulate; accordingly, knowing this mechanism is of profound and great significance. Also clearly enough, the key to showing the existence of this mechanism lies with revealing it.

The new theory (introduced in the last chapter, which is mechanism-revealed scales relativity theory) has revealed the mechanism behind the first postulate of special relativity as follows. Because velocity scale is the ratio of length scale to time scale; because this new theory shows that the length scale and the time scale in the situation of moving at a certain high speed are both reduced <u>at the same rate</u> with respect to the scales in the situation of not moving or moving at a lower speed, velocity scale is thereby constant, referred to as the constancy of velocity scale (also referred to as the constancy of speed scale), being the first basic principle of this new theory. For example, let us say that in the situation of not moving, the velocity scale is L / T (L and T are respectively the length scale and time scale in this situation), then in the situation of moving at 60 percent of the speed of light, the velocity scale is $(0.8L) / (0.8T) = L / T$, thus these two velocity scales are equal to each other. This is a bit like that $(ax)/(ay) = x/y$. Therefore, velocity scale, because it is the same at each of different speeds, is the same at different speeds.

Because velocity scale is the same at different speeds, the speed of light, due to being determined with velocity scale, is always the same (at different speeds). As a result, the speed of light is always the same, regardless of whether an observer is moving or not, regardless of whether a light source is moving or not, and regardless of whether both observer and light source are moving or not (that is, the speed of light is always the same for all observers, regardless of their motion relative to the source of light, and regardless of the relative motion between observers and light source). Thus, it is the constancy of velocity scale (or the constancy of speed scale) that determines 'the constancy of the speed of light'; or 'the constancy of the speed of light' lies with the constancy of speed scale (that is to say, behind the phenomenon or effect of 'the constancy of the speed of light' is or stands the mechanism or cause of <u>the constancy of velocity scale</u> or <u>the constancy of speed scale</u>). Therefore, **the mechanism behind the first postulate of special relativity is the constancy of velocity scale** or the constancy of speed scale (this mechanism shows that the first postulate of special relativity turns out to be a special case of the constancy of velocity scale, which can substantially help one comprehend this mechanism clearly, easily and impressively, because this postulate is so famous that many people have known it or at least heard of it). (Commentator: so, the mechanism behind the first postulate of special relativity has eventually come to light; we have finally unraveled the long-standing puzzle of *why* the speed of light is the same to every observer, no matter how he is moving. And so, from now on, this puzzle is no longer a puzzle to those who have known the answer to it. Of course, the readers of this book can congratulate themselves on having known the answer to the big puzzle that has lasted for more than one century.)

*Sincere congratulations! Dear readers, if you are professional people in physics, especially in modern physics, please accept my sincere and rational congratulations, because you have an obvious advantage over others in perceiving and realizing the (newly revealed) mechanism behind the first postulate of special relativity. The obvious advantage is that you know this postulate quite well, and you might have pondered over or asked the question of whether there is the mechanism behind this postulate, which can enable you to understand this (newly revealed) mechanism much more easily and quickly than others.

What should be pointed out or noticed is that the revealing of the mechanism behind the first postulate of special relativity is also funda-mentally important to the new theory (which is mechanism-revealed scales relativity theory, having been introduced in the last chapter) that shows us *why* time runs slower and *why* length becomes shorter at high speed (by revealing the mechanism behind these two *whys*). This fundamental importance is explicitly reflected in such a clear fact: this new theory has been verified or confirmed through its revealing the mechanism behind this fundamentally and crucially important, also exceptionally famous and highly influential, postulate, because this postulate has already become a fully recognized fact in science. Quite obviously, also rather rationally, this verification or confirmation can enable one to see the validity and reliability of this new theory from a fundamentally important angle.

Having witnessed the mechanism presented above, one can clearly realize: indeed, behind the first postulate of special relativity, there is this solid mechanism! (This is a bit like: having found out the continent of North America is the irrefutable, also undeniable, fact that there is this continent.) With such a clear realization, it seems quite reasonable and fair that one can naturally think of or ask a closely related question like: is special relativity able to reveal the mechanism behind this postulate?

The answer to this question is as clear as day: special relativity is unable to reveal the mechanism behind this postulate; that is, special relativity is utterly unable to know the underlying reason or cause behind this postulate, because this inability, believe it or not, is simply a self-evident or actually admitted *fact*. One can clearly and easily sense and grasp this *fact* via such a simple and straightforward thinking: if the mechanism behind a postulate had been revealed, the postulate would no longer have been a postulate at all; along with the constant and specific reminder from such an unavoidable *reality:* within the paradigm or stereotype of special relativity, its first postulate is <u>always</u> a postulate! Therefore, this inability of special relativity, being a self-evident or actually admitted *fact* in essence, is certainly an irrefutable or undeniable *fact* in truth. Moreover, such an irrefutable or undeniable *fact* is further witnessed or manifested after the mechanism behind this postulate has been revealed with a new theory.

What should be mentioned is that the newly developed and verified theory introduced in the last chapter, which is mechanism-revealed scales relativity theory (MRSRT, for short), is the *only* scientific theory that has the ability to reveal the mechanism behind the first postulate of special relativity, because MRSRT, as seen in chapter five, is the *only* scientific theory that is able to unveil *why* time runs slower and *why* length becomes shorter at high speed by revealing the mechanism behind these two *whys* or by revealing *why* and *how* the scales of time and length are reduced at high speed.

After introducing the mechanism behind the first postulate of special relativity, let us go to the subject of *why* and *how* the mechanism behind the second postulate of special relativity has been revealed, because it is this mechanism that tells us the *why* or cause behind this greatly important postulate (i.e., this mechanism is the reason or cause behind this postulate).

The second postulate of special relativity says that all observers moving at constant speed should have the same physical laws (that is, according to such a postulate, the laws of science should be the same for all freely moving observers, no matter what their speed). This postulate is one of the most important postulates in modern physics, also being known as one of the most famous and influential postulates in modern physics, because it is indispensable to the theory of special relativity; because of the fundamentally important status of special relativity in modern physics, as mentioned above. Therefore, if there is the mechanism behind this postulate, this mechanism is of fundamental and essential importance; accordingly, knowing this mechanism is of profound implications and substantial significance. Undoubtedly, the key to showing the existence of this mechanism lies with revealing it.

The new theory (which is mechanism-revealed scales relativity theory, introduced in chapter five) has revealed the mechanism behind the second postulate of special relativity as follows. On one side, as presented in the last chapter, this new theory shows that the scales of mass, length and time in the situation of moving at a certain high speed are <u>all</u> reduced <u>at the same rate</u> with respect to the scales in the situation of not moving or moving at a lower speed. As a result, the scale ratio of mass, length and time (which is the ratio of mass scale, length scale and time scale) is the same in all (different) situations, referred to as the constant scale ratio of mass, length and time, being the second basic principle of this new theory. For instance, if the scale ratio of mass, length and time is $M : L : T$ in the situation of not moving (M, L, and T are respectively the mass scale, length scale, and time scale in this situation), then in the situation of moving at 80 percent of the speed of light, the scale ratio is $(0.6M) : (0.6L) : (0.6T) = M : L : T$, these two scale ratios are thus equal to each other.

On the other side, in physics or science there are two types of quantities. One type is the fundamental physical quantities, which include seven quantities (they are length, mass, time, electric current, temperature, amount of substance, and luminous intensity); another type is *derived*, composite quantities (such as velocity, acceleration and force) that are derived from the related fundamental physical quantities. Therefore, the laws of physics are ultimately attributed to describing the relationships among these fundamental physical quantities. Since length, time and mass are all three fundamental physical quantities with the same fundamental status, the *prerequisite* for the second postulate of special relativity to be tenable (or the very basic requirement for this postulate to be tenable) is: the scales of mass, length and time in the state of moving at a certain high speed either all decrease or all increase with the same rate (or all remain unchanged) with respect to the scales in the state of not moving or moving at a lower speed. As a result, this *prerequisite* (or this very basic requirement) literally becomes the scale ratio of mass, length and time is the same in all (different) states, or simply becomes the constant scale ratio of mass, length and time. (What should be mentioned that the other four fundamental physical quantities are either not related to the topics that special relativity works on or not related to the speed of motion.)

When these two sides come together, and are considered simulta-neously, evidently or inevitably appears such a clear and definite conclusion: it is the constant scale ratio of mass, length and time that determines 'all observers moving at constant speed should have the same physical laws'; or it is the constant scale ratio of mass, length and time that determines 'the laws of science should be the same for all freely moving observers, no matter what their speed' (that is to say, behind the phenomenon or effect of 'all observers moving at constant speed should have the same physical laws' or 'the laws of science should be the same for all freely moving observers, no matter what their speed' is or stands the mechanism or cause of the constant scale ratio of mass, length and time). Therefore, **the mechanism behind the**

second postulate of special relativity is the constant scale ratio of mass, length and time (this mechanism explicitly points to that the second postulate of special relativity turns out to be a direct result of the constant scale ratio of mass, length and time, or a special case that comes from this constant scale ratio, to be much simpler. Such a feature can considerably help one understand this mechanism clearly, easily and impressively, because this widely recognized, famous postulate is known to be one of the most important, most prominent, most influential and most impressive postulates in modern physics). (Related question and answer: what is the key to perceiving and grasping the fact that this new theory (which is mechanism-revealed scales relativity theory, having been introduced in chapter five) has revealed the mechanism behind the second postulate of special relativity? Answer: the *prerequisite* or the least requirement for 'all observers moving at constant speed should have the same physical laws' is or lies with the constant scale ratio of mass, length and time, being the second basic principle of this new theory.)

The mechanism presented above clearly shows that really and surely, there is this solid mechanism behind the second postulate of special relativity; that is, this very postulate does have this solid mechanism! And so, the clear and definite existence of this solid mechanism turns out to be an irrefutable fact. This is quite similar to: no rational people in the world are doubtful about the existence of the continent of North America, after having found out this continent. This is also a bit like: having found out diamond beneath a certain place is the solid evidence that there is diamond beneath this place. (Commentator: yes, one can clearly see this solid mechanism, simply because it has been revealed.)

*Sincere congratulations from the author! Dear readers, if you are the professional people in physics, especially in modern physics, please accept my sincere and rational congratulations, because you have an obvious advantage over others in comprehending the (newly

41

revealed) mechanism behind the second postulate of special relativity. The obvious advantage is that you know this postulate pretty well, and you might have pondered or asked the question of whether there is the mechanism behind this postulate, which can help or make you understand this (newly revealed) mechanism much more easily and quickly than others.

What should be mentioned is that the revealing of the mechanism behind the second postulate of special relativity is also crucially important to the new theory that unveils *why* time runs slower and *why* length becomes shorter at high speed by revealing the mechanism behind these two *whys* (again, this new theory is mechanism-revealed scales relativity theory, which has been introduced in chapter five). This crucial importance is clearly demonstrated or reflected in such an explicit fact: this new theory has been verified or confirmed through its revealing the mechanism behind this crucially important, also highly famous and influential, postulate, because this postulate has been fully recognized or widely acknowledged in science for quite a long time. Quite reasonably, also very perceptibly, this verification or confirmation can enable one to perceive and realize the validity and reliability of this new theory from a crucially important angle. As a result, this new theory has been verified or confirmed from three different aspects or angles altogether; through these comprehensive verifications or confirmations, the validity and reliability of this new theory have been (or could be) clearly seen from different angles. (*Friendly reminder: altogether the new theory that unveils *why* time runs slower and *why* length becomes shorter at high speed has been verified or confirmed from the following three aspects or angles. First, this new theory has been verified or confirmed from the angle of its fundamental theoretical foundation— the law of mass doing work, via its revealing the mechanism of/behind the famous mass-energy equivalence equation, as shown in chapter three. Second, this new theory has been verified or confirmed through its revealing the mechanism behind the first postulate of special relativity, as introduced earlier in this chapter. Third, this new theory has been verified or confirmed through its revealing the mechanism behind the second postulate of special relativity, as just presented above.)

What should be pointed out or noticed is that the revealing of the mechanism behind the second postulate of special relativity has profound implications. One of them inevitably points to: special relativity is unable to reveal the mechanism behind its second postulate (that is, special relativity is unable to dig out the underlying reason or cause behind this postulate), simply because this inability, believe it or not, is a self-evident or actually admitted *fact*. One can clearly and easily perceive and realize this *fact* via such a simple and clear thinking: if the mechanism behind a postulate had been revealed, the postulate would no longer have been a postulate at all. In other words, the unavoidable *reality*, which is that this postulate is <u>always</u> a postulate within the paradigm or stereotype of special relativity, is exactly the hard evidence or clear manifestation that special relativity is unable to reveal the mechanism behind this postulate. So this inability of special relativity is indeed a self-evident or actually admitted *fact;* such an inability is, of course, also an undeniable or irrefutable *fact* in truth. More noticeably, this inability is further witnessed after the mechanism behind this postulate has been revealed with a new theory.

What should be mentioned is that the newly developed and verified theory introduced in chapter five, which is mechanism-revealed scales relativity theory, is the <u>only</u> scientific theory that has the ability to reveal the mechanism behind the second postulate of special relativity (or that has the ability to find out the reason or cause behind this postulate), because this new theory, and <u>only</u> this new theory, is able to reveal and prove the constant scale ratio of mass, length and time, being the second basic principle of this new theory, as presented above;

because the mechanism behind this postulate is the constant scale ratio of mass, length and time, as analyzed and shown above.

*Commentator: after revealing the mechanism behind the two postulates of special relativity, what should we do? We should genuinely and enthusiastically congratulate these two fundamentally and crucially important, tremendously influential and exceptionally famous postulates on this unprecedented revealing with profound implications and historic significance. The theory of special relativity should congratulate itself on such an unprecedented revealing, and should celebrate such an unprecedented revealing. The professional people working on or in the area of special relativity should sincerely and zealously congratulate special relativity on such an unprecedented revealing, and should actively and joyfully celebrate such an unprecedented revealing. (The response from the author: the above comments also represent my voice. More than that, the readers of this book should genuinely and delightedly congratulate themselves on having known the mechanism behind these two greatly important and extraordinarily famous postulates.)

Chapter 7

The Grave Consequences of Not Knowing Why the Speed of Light Is Constant

(The highlight or climax of this chapter)

The undeniable basic fact, which is that special relativity doesn't have the ability to know *why* the speed of light is constant, has resulted in the hard fact that special relativity is unable to know *why* time runs slower and *why* length becomes shorter at high speed. This hard fact has, in turn, led to the unavoidable fact that special relativity turns out to be clearly and seriously self-contradictory, because its two <u>core</u> concepts (time dilation and length contraction) turn out to be clearly and seriously self-contradictory, as revealed by the hard evidence shown in this chapter. The second evidence that strongly supports this unavoidable fact is: the famous Twin Paradox, when considered together with relativistic mass (mass increase with speed) and length contraction—the Twin Paradox, relativistic mass and length contraction are <u>all</u> the three products of special relativity, turns out to be clearly, even obviously, a paradox. The third hard evidence that strongly supports this unavoidable fact is: the five <u>main</u> components of special relativity (time dilation, length contraction, relativistic mass, and its two postulates) cannot be put into the same package of special relativity at all.

Many people, needless to say the professional people in physics, have clearly known the (undeniable) basic fact that the theory of special relativity doesn't have the ability to answer the long-standing, big question of *why* the speed of light is constant, being the same thing as that this theory is unable to solve the long-standing, big problem of *why* the speed of light is constant. That is, this basic fact is that special relativity doesn't have the ability to know *why* the speed of light is constant.

What could this basic fact enable one to think of or think over then? It seems rather rational and quite reasonable (at least neither irrational nor unreasonable) that this basic fact could enable one to think of or think over thus ask a closely related, also greatly important, question like: has this basic fact resulted in or caused serious or critical outcomes?

The answer to the above question is clear and definite: this basic fact has indeed resulted in or caused serious outcomes; and the most serious outcome is such a hard fact: special relativity doesn't have the ability to solve the fundamental problems of *why* time runs slower and *why* length becomes shorter at high speed, being the same thing as that special relativity doesn't have the ability to answer the fundamental questions of *why* time runs slower and *why* length becomes shorter at high speed (related reminder: for the detailed, specific, penetrating and comprehensive information about this hard fact, please go back to chapter five if necessary). (Narrator: that is to say, special relativity, because it is unable to answer the question or solve the problem of *why* the speed of light is constant, doesn't have the ability to answer the questions or solve the problems of *why* time runs slower and *why* length becomes shorter at high speed.)

Then what could this hard fact make or help one think of or think over? It is rather rational and quite reasonable (at least neither irrational nor unreasonable) that this hard fact could make or help one think of or think over thus raise closely related, also fundamentally and crucially important, questions like: has this hard fact resulted in or caused grave, even disastrous, consequences? If the answer is yes, what are these consequences then?

The task of this chapter is to find out the gravest, also extremely disastrous, consequence that has been resulted in or caused by this hard fact. The specific method to carry out this task is to inspect whether special relativity turns out to be clearly and seriously self-contradictory.

How do we inspect whether special relativity turns out to be clearly and seriously self-contradictory then? Answer: we need to inspect whether its <u>core</u> concepts are clearly and seriously self-contradictory. What are its <u>core</u> concepts then? Answer: the <u>core</u> concepts of special relativity are length contraction and time dilation, respectively for interpreting length becomes shorter and time runs slower that appear in the situation of high speed. In many materials on special relativity, length contraction and time dilation are very often said to be the most crucial core or innermost core of special relativity; and some of them even go further by saying that length contraction and time dilation are more or less equivalent to special relativity. Therefore, in order to inspect whether special relativity turns out to be clearly and seriously self-contradictory, what we need to do is to examine or scrutinize whether length contraction and time dilation turn out to be clearly and seriously self-contradictory. If length contraction and time dilation turn out to be factually incompatible thus essentially contradictory with each other, then they are clearly and seriously self-contradictory; that is, special relativity turns out to be clearly and seriously self-contradictory.

Now let us examine whether length contraction and time dilation are factually compatible or not. At present, in describing length contraction and time dilation, they are factually separate from each other! For instance, in demonstrating length contraction, the following scheme is often involved (Fig. 7.1, P. 49). An object, which can be a spaceship (say, its rest length is 250 feet and its height is 50 feet), approaches a 50-foot vertical line when its velocity is very close to the speed of light, because the original 250-foot rest length is shortened close to zero by length contraction. (If there is a meter stick in <u>the</u> spaceship, this meter stick is shortened close to zero either by length contraction, because it must be placed along the length direction of the spaceship for measuring its length.) On the other hand, in illustrating time dilation, three identical clocks are used: two are at rest in a stationary spaceship, and one is in a moving spaceship (Fig. 7.2, next page). What should be well aware is that the clock in the moving spaceship is always identical to the two at rest, even when the velocity of the spaceship is increased up to close to the speed of light (this spaceship is thus forced to throw away its length contraction utterly: it does not experience length contraction at all). (One can easily get an animated description of length contraction and time dilation as mentioned above, from today's web with key words such as 'length contraction' and 'time dilation' etc.) Please closely notice what I want to emphasize and point out: the clock in the spaceship moving at high speed **does not** experience length contraction at all! And moreover, in the various textbooks on special relativity, the clock moving at high speed **never** experiences length contraction either (thus many professional people in physics must or might have already perceived and noticed this obvious feature!). Here a crucially important question is directly coming towards us: can a meter stick, which is used to measure length or space, be put together with a clock in the same spaceship moving at high speed? If not, why? The answer will be seen in the coming two paragraphs.

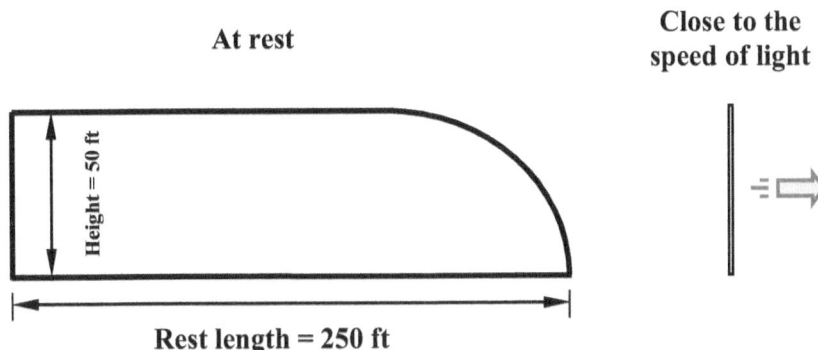

Figure 7.1, the length contraction in the theory of special relativity, which is responsible for interpreting the phenomenon of length becoming shorter at high speed.

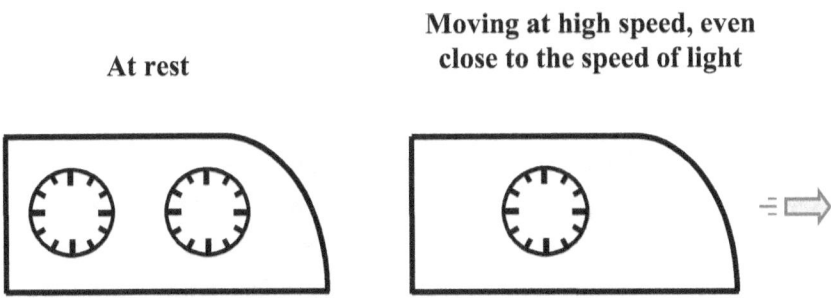

Figure 7.2, the time dilation in the theory of special relativity, which is responsible for interpreting the phenomenon of time running slower at high speed.

What would happen if the description of length contraction and time dilation were presented in the manner that a meter stick and a clock are put in the <u>same</u> spaceship moving at high speed? One of the following three obviously and seriously contradictory situations would definitely appear within the paradigm or stereotype of special relativity (Fig. 7.3, next page)! <u>Situation 1</u>: according to the description above, the meter stick would experience length contraction (in the same way as the spaceship illustrated in Fig. 7.1, because the meter stick has to be placed along the length direction of the spaceship in order to measure its length); whereas the clock, which is certainly made up of a certain amount of mass—just like the meter stick, would not experience length contraction at all. This is clearly self-contradictory. <u>Situation 2</u>: admitting the clock would experience length contraction (for being consistent with the meter stick), then this clock would be squeezed into "0" or even "l" shape from the original "O" shape by its length contraction. This is clearly and directly contradictory with what special relativity has said about time running slower via its time dilation—the clock moving at high speed **never** experiences length contraction at all in special relativity (even if temporarily putting aside the issue of whether such a seriously squeezed clock could still work or not. Of course, it seems rather rational if one has the reasoning: most likely, such a seriously squeezed clock couldn't work, because it would be very skeptical that this clock could still be able to measure time accurately. That is, such reasoning seems to be reasonably acceptable to most people, except those who want to refuse it on purpose with some extreme excuses). <u>Situation 3</u>: denying the clock would experience length contraction (for being consistent with the requirement of special relativity), then the meter stick could/would not experience length contraction either. This is clearly incompatible or plainly contradictory with another requirement of special relativity: a meter stick must experience length contraction at high speed.

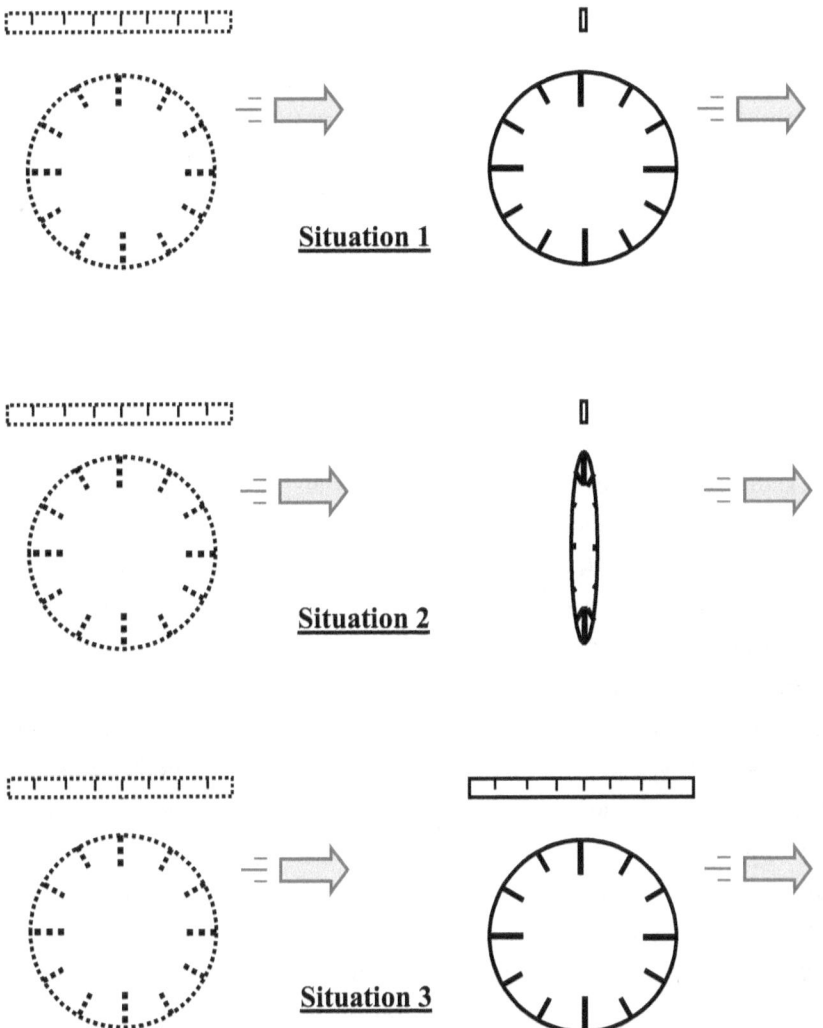

Figure 7.3, a meter stick and a clock cannot be put together in the same spaceship moving at high speed, according to the theory of special relativity.

So each of these three situations shows that special relativity has caused obviously ridiculous results! In fact, only these three situations could occur within the paradigm of special relativity. Moreover, what ought to be pointed out and noticed is: meter stick and clock are **never** put together either in the various textbooks on special relativity when its length contraction and time dilation are interpreted; that is, many insightful professional people in physics, when teaching and studying the theory of special relativity, must or might have already realized such a noticeable feature. We have thus seen a crucially important *reality*—a clear *reality*, an unavoidable *reality*, also an undeniable *reality*, which is that within the paradigm of special relativity, a meter stick and a clock cannot be put together in the <u>same</u> spaceship moving at high speed. (The focus or crux of this *reality* is: according to the theory of special relativity, a meter stick must experience length contraction in the state of high speed, whereas a clock must **NOT** experience length contraction in such a state.)

Then what does such a *reality* really tell us? Or what are the actual implications of such a *reality?* <u>Clearly and obviously</u>, length contraction and time dilation (respectively for interpreting that length becomes shorter and time runs slower, the two phenomena that appear in the situation of high speed, as just mentioned above) are factually incompatible thus essentially contradictory, because they would directly deny each other if met together. <u>Evidently and undeniably</u>, length contraction and time dilation, therefore, turn out to be clearly and seriously self-contradictory. (Narrator: stated plainly, this clear and serious self-contradiction is explicitly reflected in the unavoidable, unconcealed *fact* that length contraction and time dilation don't and can't coexist at all! This is because: in order to demonstrate length contraction with a spaceship moving at high speed, a clock, which is used to measure time thus to show time dilation, is utterly not allowed to be put in *the very* spaceship; in order to illustrate time dilation, the length contraction of a spaceship moving at high speed is forcibly given up!) <u>Inescapably and inevitably</u>, special relativity turns out to be a theory that is clearly and seriously self-contradictory, because its two <u>core</u> or <u>key</u> concepts, length contraction and time dilation, turn out to be clearly and seriously self-contradictory. (Commentator: in science, any theory, including special relativity of course, as long as its two <u>core</u> or <u>key</u> concepts turn out to be clearly and seriously self-contradictory, the theory is clearly and seriously self-contradictory, being a plain rationale or self-evident truth, thus also being a clear and irrefutable rule. In fact, this rule is so obvious that there is no need to explain anything about it at all. In other words, it is not only obviously rational but also definitely sensible to believe that no real scientists would deny such an irrefutable rule. Moreover, if anybody disagreed with this irrefutable rule in the name of a scientist, he or she would become a big laughingstock of all rational people; in reality, it seems clear enough that no real scientists want to deny or challenge this irrefutable rule.)

So, succinctly and accurately, the *reality* revealed above (which is that special relativity utterly does not allow a meter stick and a clock to be put together in the <u>same</u> spaceship moving at high speed) clearly, definitely and irrefutably shows the three inherently connected <u>facts</u> as follows. Length contraction and time dilation are factually incompatible thus essentially contradictory with each other (a plain fact) → they are clearly and seriously self-contradictory (a solid fact) → special relativity turns out to be clearly and seriously self-contradictory (an unavoidable fact).

*Related questions and answers: what is the relationship of the above three facts? Answer: from the plain fact to the solid fact to the unavoidable fact is inescapable or inevitable, thus the only route. Thus, once one has fully understood, through the hard evidence displayed above, the plain fact, he or she will have no difficulty grasping the solid fact and the unavoidable fact. Question: then how could one see, perceive and realize the plain fact (which is that length contraction and time dilation are factually incompatible thus essentially contradictory with each other) even more clearly and explicitly, thus grasp it more tangibly and impressively, and sense it more definitely and confidently from other noticeable angles, besides the irrefutable or hard evidence presented above? Answer: the specific information in the coming paragraph can provide an explicit and substantial help.

The plain fact, which is that length contraction and time dilation are factually incompatible thus essentially contradictory with each other, becomes crystal clear and transparently explicit through the intense dispute between a spaceship about to travel at high speed and a clock inside it. (The spaceship represents length contraction, and the clock stands for time dilation in their dispute.) Before leaving, the spaceship started first: I would experience length contraction according to special relativity, and would approach a vertical line as my original height when my velocity is close to the speed of light. The clock refuted seriously: no, no, in that situation I would be squeezed into "0" or even "1" shape by your length contraction from my original "O" shape; this would put me in a position that is directly contradictory with the requirement of special relativity on me—according to special relativity, I must NOT experience length contraction at high speed. Besides, I am afraid this would make me unable to measure time. So let me say first: I want to work properly and measure time accurately, according to the requirement of special relativity on me—that is, I will NOT experience length contraction. The spaceship argued without delay: no, no, in that situation I would NOT experience length contraction at all, as required on me by special relativity. Being upset and angry, the spaceship yelled to the clock: get off me; if not, you will destroy my length contraction! Being badly annoyed and hurt, the clock cried: of course, I will get off, because otherwise your length contraction will make me unable to measure time, and … and I will become useless, and … and worthless!

(*Book reviewer: quite obviously, through the intense dispute above, the plain fact, which is that length contraction and time dilation are factually incompatible thus essentially contradictory with each other, manifests itself even further, thus becomes much more noticeable; such an effect is, of course, substantially helpful for one to realize this plain fact even more clearly and explicitly. This realization, in turn, can enable one to be well aware of the solid fact that length contraction and time dilation turn out to be clearly and seriously self-contradictory. With this solid fact kept in mind, it seems clear enough or rather rational that one can clearly understand the unavoidable fact that special relativity turns out to be clearly and seriously self-contradictory indeed, because length contraction and time dilation are its two <u>core</u> or <u>key</u> concepts, as pointed out above.)

On seeing the solid *fact* that length contraction and time dilation (respectively for interpreting length becomes shorter and time runs slower at high speed) turn out to be clearly and seriously self-contradictory, one might be bewildered: it is often said that special relativity has passed observational tests, why does this solid fact still exist? Regarding the possible bewilderment of this sort, please allow me to clarify it as follows. This clarification is: all observational tests of special relativity have neither the function nor the ability to change or affect this solid *fact* at all (this is a bit like that all observational and experimental tests of general relativity and quantum mechanics still cannot change or affect the well-known *fact* that these two theories are seriously inconsistent or contradictory with each other). As a matter of fact, all observational tests of special relativity have never ever targeted to deal with this solid fact; not to mention that none of these tests has even intended to touch this solid fact at all! That is to say, all these observational tests actually have nothing to do with this solid fact, believe it or not (consequently, this solid fact still exists).

About this solid *fact*, another related clarification is also provided as follows. While separately and individually, and viewing length contraction and time dilation one by one, it looks as if they are compatible or consistent, just because their incompatibility or inconsistency doesn't emerge when they are viewed separately and individually. However, when length contraction and time dilation are pieced together, and are considered simultaneously, they are (really and truly) factually incompatible thus directly contradictory with each other—that is, they are clearly and seriously self-contradictory indeed.

(*Related questions and answers: what bitter lesson can we draw or learn from the solid fact that length contraction and time dilation turn out to be clearly and seriously self-contradictory? Answer: comprehensive and careful inspection of the different aspects of a theory is fundamentally important in science, if the theory deals with two or more different aspects. Question: what is the crux of this solid fact then? Answer: length contraction cannot tell *why* length becomes shorter at high speed, and time dilation cannot tell *why* time runs slower at high speed, being the most serious outcome of the basic fact that the theory of special relativity doesn't have the ability to know *why* the speed of light is constant. Moreover, this answer becomes even clearer through a sharp contrast against the new theory that has uncovered the secret mask of *why* length becomes shorter and *why* time runs slower at high speed; this new theory has been introduced in chapter five.)

After seeing the (newly revealed) unavoidable fact that special relativity turns out to be clearly and seriously self-contradictory (because its two <u>core</u> or <u>key</u> concepts—length contraction and time dilation turn out to be clearly and seriously self-contradictory, or because of the solid fact revealed and shown above), dear readers might have two typical or likely reactions. Some readers perhaps want to know more evidence about this unavoidable fact; others might feel a big surprise (probably because this unavoidable fact is too new or too radical in their eyes, along with it appearing too suddenly in their minds). Either of these possible reactions seems to call for a verification of this unavoidable fact.

How do we verify this unavoidable fact then? Answer: the verification of this unavoidable fact is to be carried out by inspecting whether the famous Twin Paradox is truly a paradox or not in the coming several paragraphs (when a concept, an idea, or a story is self-contradictory, we often refer to it as a paradox. The famous Twin Paradox is a highly influential story not only in special relativity but also in science fiction). If this Paradox turns out to be truly a paradox, then this unavoidable fact is verified from a remarkably important angle. On the contrary, if it turns out that this Paradox is not a paradox, then one has a remarkably important reason to doubt, even deny, this unavoidable fact. So the basis or rule of judging this verification is objective, rational, clear and fair.

Well, as a preparation, let us review the famous Twin Paradox at the start. The Twin Paradox arises from the effect of time dilation (which is the effect of time running slower in the situation of high speed, because in special relativity, time dilation is responsible for interpreting the phenomenon of time running slower in such a situation, as pointed out earlier). It says there are twin brothers. One brother stays at home in a place of a slow speed (on the earth, for example). The other brother goes away for a long trip in a very fast spaceship. The "slow" twin ages considerably; he has lost most of his hair. The younger twin returns still young; he still has all of his hair. This imagined situation is called the famous Twin Paradox. (The story of the Twin Paradox appears in many materials that talk about or explain the theory of special relativity. A lot of people might have already known this prevalent and famous story from various sources.)

In order to check whether the famous Twin Paradox is truly a para-dox or not, another necessary preparation is to know about the concept or idea of relativistic mass (that is, mass increase with speed), an important concept formed within the paradigm of special relativity. This concept says that: the mass of an object increases with the increase in its velocity, and the mass of an object becomes infinitely large when the object infinitely approaches the speed of light. So 'relativistic mass' is often simply said as 'rest mass is least'. The name of relativistic mass has frequently appeared in the various materials on special relativity. (*Note: the concept or idea of relativistic mass has also been mentioned in chapter four as an important issue for avoiding the possible confusion between relativistic mass and the mass consumption.)

Now we are ready to check whether the famous Twin Paradox is truly a paradox or not. How will this be done? It will be checked together with relativistic mass (mass increase with speed) that comes from special relativity, just like the Twin Paradox. Let us say, each of the twin brothers has the rest mass of 75 kilograms (165 pounds), and the spaceship, in which one brother of the twins would stay, would be traveling at the velocity of 98 percent of the speed of light. Now let us see what would happen to the brother traveling away.

According to relativistic mass, the mass of the brother in the spaceship would increase from 75 to 377 kilograms (165 to 830 pounds)! How could this dear brother still be alive after experiencing such an explosive increase of mass?!? Clearly, no rational people can think this result is reasonable. When he returned from this trip—if he could still be alive, his mass would shrink from 377 to 75 kilograms (830 to 165 pounds)! How could this dear brother still be alive after experiencing such a rapid decrease of mass like a quick evaporation?!? Also clearly, no one with clear thinking can imagine this result is valid. Therefore, the Twin Paradox, when considered together with relativistic mass (mass increase with speed), turns out to be a self-contradictory story. That is, the famous Twin Paradox turns out to be truly a paradox as long as it is considered simultaneously with relativistic mass; and no excuses or "explanations" can change such a clear and definite fate. (Related question and answer: your analysis and conclusion above might cause an argument like: no, no, the Twin Paradox is just for explaining the effect of time dilation, so you shouldn't put this Paradox together with relativistic mass. What is your response to that sort of possible argument? Answer: the very *fact*, which is that the Twin Paradox and relativistic mass couldn't be put together, actually and exactly shows that special relativity turns out to be clearly and seriously self-contradictory. Please notice that both the Twin Paradox and relativistic mass are the products of special relativity!)

In fact, when considering together with length contraction (which is responsible for interpreting the phenomenon of length becoming shorter at high speed), one can see even more clearly that the famous Twin Paradox turns out to be really and truly a paradox. For the brother in the spaceship traveling at the velocity of 98 percent of the speed of light, his mass would increase from 75 to 377 kilograms, thus he would be 5.03 times his original mass (this would enlarge his area from 1 to 5.03), but his length would only be 1/5.03 of his original length (this factor would also expand his area by the ratio of 1 to 5.03). So the area of this brother would be about 25 times his original size; this dear, but poor, brother would thus be squeezed into a big but very thin board, like a large piece of plywood.

Now, one can clearly see that the Twin Paradox, when considered together with relativistic mass (mass increase with speed) and length contraction, turns out to be a truly self-contradictory story indeed. Moreover, neither excuses/pretexts nor "explanations" have the magic power to change this definite fate! Consequently, also unavoidably and undeniably, the very *fact*, which is that the Twin Paradox, relativistic mass and length contraction cannot be put together at all—they three are utterly not compatible, is a clear reflection or concrete embodiment that special relativity turns out to be a theory that is clearly and truly self-contradictory. (Commentator: now, it becomes clear that the prevalent opinion, which says that the Twin Paradox seems not paradoxical as long as one abandons the idea of absolute time, turns out to be simply the consequence of merely staring at the effect of time dilation—the effect of time running slower at high speed, while utterly forgetting relativistic mass and length contraction. Of course, such an opinion turns out to be neither valid nor correct; please remember that the Twin Paradox, relativistic mass and length contraction are all the three products of special relativity.)

Well, people will see the new story of the famous Twin Paradox is quite different from the original one. I think we can rewrite this new story using the results just revealed above. Let us say, in the United

States there was a senior professor in physics, also an expert on special relativity, whose name was George. George had two twin sons, John and Mark. They almost finished their study for a master degree, and planned to continue their education for a Ph.D. degree. On a Friday evening, they three were chatting around a table after dinner.

George started first, "Since both of your majors are physics, I am sure you guys are familiar with the famous story of the Twin Paradox."

"Yes, we are," John and Mark answered quickly.

"Then let's talk about the feasibility of testing the famous Twin Paradox, just a simulated test of course," George continued.

"Ok," the twin sons replied without hesitation.

"Let's say Mark will be the brother about to go away for a long trip in a spaceship moving at 99 percent of the speed of light, and John will be the brother to stay at home on the earth," George assigned the roles to his twin sons.

"I need some time to think over my role," Mark said. "I will also think about the role of Mark, because my role is so simple that there is no need to be worried about it," John remarked.

"Go ahead, guys," George agreed gladly.

About one month later, their discussion came back. This time Mark began first, "I've found a big problem in the famous story of the Twin Paradox. According to the idea of mass increase with speed, that is, the so-called concept of relativistic mass, my mass would be increased up to 503 kilograms at 99 percent of the speed of light from the present 71 kilograms. You know, 503 kilograms is more than 1100 pounds." "Yes, I've also seen the same problem," John agreed quickly.

The result from his twin sons was really a great surprise to George, so he recalculated. And he also got the same result as Mark and John! The great surprise, however, still continued hanging around George. He kept asking himself: why, why … and why I have not thought of the issue of mass increase with speed, I should have noticed that earlier … I should have thought of that before.

But an even greater surprise to George was the second result of Mark. After pausing for a moment, Mark continued, "Moreover, from the idea of length contraction embedded in special relativity, my length, which is actually my front-to-back thickness, suppose I would keep upright in the spaceship and face the direction in which it is moving, would be only about 1/7.1 of my original thickness. This factor would also expand my area by the ratio of 1 to 7.1, plus the idea of mass increase with speed would enlarge my area from 1 to 7.1, thus my face-to-face area would be about 50 times the original area. So I would virtually become a large picture." "Oh, really?" John also felt a bit of surprise this time. But his recalculation showed that Mark's result was correct indeed! After George checked the result of his two sons, though correct, he felt the surprise was still so big that he couldn't face it.

For a couple weeks, George kept asking and thinking, "Why have I not considered time dilation, mass increase with speed, and length contraction together, why, why, and why? I should have thought of these three things together, I should have...." In that kind of mood, poor George was not happy at all, of course. His wife, Jennifer, certainly noticed the sadness of her dear George; she also knew why he was so unhappy after asking her twin sons. She decided to help her dear and poor George, but didn't know how. Eventually, or by accident, Jennifer told George the fable of 'the Blind Men and the Elephant'. After George heard this wonderful fable, being suddenly enlightened, he became happy again. This fable seemed to have an amazing effect on George!

Then what is the fable of 'the Blind Men and the Elephant'? Is it a miraculous panacea? I couldn't help sharing it with dear readers here. Once upon a time, there were six blind men who lived in a village in today's India. Every day they stood begging on the side of the road. They had often heard of elephants, but had never seen one, for being blind, how could they? One morning, it happened that an elephant was led down the road at which they exactly stood. When they heard that an elephant was passing by, they asked the drover to stop the beast so that they could have a "look". Of course they could not look at him with their eyes, but they thought they might learn what kind of animal he was by touching and feeling him. Then you should see they trusted their own sense of touch so much.

The first blind man happened to put his hand on the side of the elephant. So he said, "Well, well, this beast is exactly like a wall."

The second tightly grasped one of the elephant's tusks and felt it. So he said loudly, "You're quite mistaken. He's round, smooth, and sharp. He's more like a spear than anything else."

The third happened to grab the elephant's trunk. "Both of you are completely wrong. This elephant is like a snake," he said.

The fourth opened both his arms and closed them around one of the elephant's legs. "Oh, how blind you are!" he cried. "It's very clear that he's round and tall like a tree."

The fifth man was very tall, so he caught one of the elephant's ears. "Even the blindest person must see that this elephant isn't like any of the things you name at all!" he said. "He's exactly like a huge fan."

The sixth man went forward to touch and feel the elephant. He was old and slow, so it took him quite some time even to find the elephant. Eventually, he got hold of the beast's tail. "Oh, how silly you all are!" cried he. "The elephant isn't like a wall, or a spear, or a snake, or a tree; neither is he like a fan. Any person with eyes on head can see that he's exactly like a rope."

After the drover and the elephant left, the six men sat by the roadside all day, quarrelling about the elephant. They could not agree with one another, because each of them so surely believed that he knew what the beast looked like. It is not merely blind men who make such stupid mistakes. People who can see sometimes may act just as foolishly.

Having shared the two stories above, let us still go back to the famous Twin Paradox and finish the discussion about this highly influential Paradox. Up to here, and at this moment, several important and noticeable questions about this Paradox are probably hovering over the heads of some, even many, readers. Most importantly or noticeably, what bitter lesson does the Twin Paradox really teach us? And what painful lesson can we draw from this Paradox? More specifically, since the Twin Paradox turns out to be clearly a paradox once relativistic mass (mass increase with speed) comes in, **why** has it been so prevalent for a long time? Since the Twin Paradox turns out to be obviously a paradox only if seen together with relativistic mass and length contraction (which is used to interpret the phenomenon of length becoming shorter at high speed), **why** has it been a fashionable example in special relativity? In other words, or more plainly, since the Twin Paradox turns out to be really and truly a paradox within the paradigm of special relativity itself (thus, one may suddenly realize that the so-called famous Twin Paradox turns out to be an 'infamous' story), **why** have some theoretical physicists been delighted to talk about it without boredom?

To these questions, different people may have different answers because different people can look at the same thing in quite different ways. But I like to share my answer with dear readers here (though I firmly believe that many insightful readers may have better answers than me). In science, if <u>several aspects</u> are involved in the same theory, one should consider and carefully examine <u>them</u> together to check whether <u>they</u> are compatible or not; these different aspects should be closely scrutinized simultaneously to inspect whether they are self-contradictory or not. That is, we ought to view <u>the different aspects</u> of the same theory as a whole, in order to make sure <u>they</u> are compatible or consistent with each other, in order to ensure <u>they</u> are not self-contradictory with one another. Otherwise, it would be highly risky—we could easily make the ridiculous mistakes like in 'the Blind Men and the Elephant', if our eyes were merely kept on one aspect while forgetting the others. Of course, it would be even more dangerous—we could easily, even inevitably, make stupid mistakes, if we simply threw away the other aspects in dealing with one aspect, once we had noticed or perceived that self-contradiction is inescapable when these aspects meet together. (Commentator: the history of science development has clearly warned or told us such a bitter or painful lesson from both positive and negative experiences: wrong thinking from wrong concept is the number one enemy of science! And it seems quite safe to say that all good or experienced scientists would definitely agree with and/or firmly believe: correct thinking is crucially and decisively important in science; correct concept is the key to developing correct thinking.)

Up to here, the task of verifying the unavoidable fact that special relativity turns out to be clearly and seriously self-contradictory has been completed by inspecting whether the famous Twin Paradox is really a paradox or not. The inescapable conclusion obtained from the completion of this task is: the Twin Paradox turns out to be clearly, even obviously, a paradox—that is, this Paradox turns out to be really and truly a paradox once relativistic mass (mass increase with speed) comes in; the Twin Paradox turns out to be obviously a paradox only if seen together with relativistic mass and length contraction. Quite obviously, such an inescapable conclusion strongly supports this unavoidable fact; as a result, this verification is definitely a considerable and impressive help for one to face and comprehend this unavoidable fact.

Even though the unavoidable fact, which is that special relativity turns out to be clearly and seriously self-contradictory, has been verified; even though this verification explicitly shows that this unavoidable fact is clearly true, and definitely true, I still feel it is better or necessary to go the extra mile about this unavoidable fact, considering such a fact might still be a big surprise to some people or quite new to them. How do we go the extra mile? Answer: it will be carried out by confirming this unavoidable fact. How do we confirm this unavoidable fact then? Because special relativity has five <u>main</u> components, the task of confirming this unavoidable fact is thus to be carried out by inspecting whether these five components are compatible or not in the following several paragraphs; that is, to check whether these five components can be put into the same package of special relativity.

The five main components of special relativity are: length contraction (for interpreting length becomes shorter at high speed), time dilation (for interpreting time runs slower at high speed), relativistic mass (mass increase with speed), and its two postulates. Thus, if these five components are not compatible—that is, if they cannot be put into the same package of special relativity (which is a further demonstration of the self-contradictory feature of special relativity from the wide angle of its entire structure), then the unavoidable fact (which is that special relativity turns out to be clearly and seriously self-contradictory) has been confirmed. On the other hand, if these five components are compatible—that is, if they can be put into the same package of special relativity, then one has a very important reason to doubt, even disagree with, this unavoidable fact. So the basis or criterion of judging the result to be obtained from carrying out this task is rational, objective, clear and fair.

As a preparation, let us have a quick look at the two postulates of special relativity. Its first postulate says that the speed of light is the same for all observers, regardless of their motion relative to the source of light; 'the constancy of the speed of light', to be simpler. That is to say, this postulate tells us that all observers should measure the same speed of light, no matter how fast they are moving. The second postulate of special relativity says that all observers moving at constant speed should have the same physical laws; that is, according to this postulate, the laws of science should be the same for all freely moving observers, no matter what their speed. (*Related reminder: these two postulates have been mentioned more specifically in the last chapter.)

Now we get ready to inspect whether the five main components of special relativity are compatible or not. Let us start by finding out the answer to the question: can length contraction, time dilation and relativistic mass be compatible with the second postulate of special relativity? As mentioned in the last chapter, in physics or science there are two types of quantities. One type is the fundamental physical quantities, which include seven quantities (they are length, mass, time, electric current, temperature, amount of substance, and luminous intensity); another type is *derived*, composite quantities (such as velocity, acceleration and force) that are derived from the related fundamental physical quantities. As a result, the laws of physics are ultimately attributed to describing the relationships among these fundamental physical quantities. Since length, time and mass are <u>all</u> three fundamental physical quantities with the <u>same</u> fundamental status, for the second postulate of special relativity to be tenable, the *prerequisite* is that the changes in length, time and mass **have to** be <u>at the same rate</u> in the state of moving at a certain high speed (that is, they three (length, time and mass) either all decrease or all increase <u>with the same rate</u> in such a state) with respect to the state of not moving or moving at a lower speed. However, the actual meanings of length contraction, time dilation and relativistic mass are: length becomes **shorter**, and time runs **slower**, whereas mass becomes **larger**. Consequently, the combination of length contraction, time dilation and relativistic mass literally takes away such a *prerequisite*, which is the minimum or least requirement for the second postulate of special relativity to be tenable. Therefore, it turns out that length contraction, time dilation and relativistic mass cannot be compatible with the second postulate of special relativity at all.

After seeing the analysis and conclusion in the paragraph above, one might argue or ask: how about if relativistic mass stands together with the second postulate of special relativity? Let us see what would happen then. If relativistic mass (mass increase with speed—that is, mass becomes larger) were put together with this postulate, and maintained its validity, then the inevitable requirement on length and time would become that length becomes **longer** and time runs **quicker**. Such a requirement directly conflicts with length contraction and time dilation, because they respectively correspond to length becoming **shorter** and time running **slower** in the situation of high speed (which, in turn, is because in special relativity, length contraction is responsible for interpreting length becomes shorter at high speed, and time dilation is responsible for interpreting time runs slower at high speed). Not only that, since length contraction and time dilation are the direct result of fitting the first postulate of special relativity, the effect of this postulate is actually reflected in length contraction and time dilation; thus they (length contraction and time dilation) represent this postulate in fact. Therefore, relativistic mass, when standing together with the second postulate of special relativity, cannot be compatible with its first postulate at all. (What should be clarified or pointed out is: the above analyses and conclusions neither say the two postulates of special relativity *themselves* are not correct, nor these two postulates themselves *alone* are not compatible.)

Through the analyses and conclusions in the two paragraphs above, one can clearly see that the five <u>main</u> components of special relativity, which are length contraction (for interpreting length becomes shorter), time dilation (for interpreting time runs slower), relativistic mass (mass increase with speed), and its two postulates, turn out to be obviously incompatible in truth, because these five <u>main</u> components cannot be put into the same package of special relativity at all. Such an obvious incompatibility is undoubtedly a further demonstration of the self-contradictory feature of special relativity from the wide angle of its entire structure. (Commentator: quite noticeably, this obvious incompatibility, being the third time showing the self-contradictory feature of special relativity, can substantially help one perceive this self-contradictory feature more comprehensively and thoroughly. This perception can enable one to realize this self-contradictory feature more clearly and definitely.) Therefore, as a reminder, the unavoidable fact (which is that special relativity turns out to be clearly and seriously self-contradictory) has been confirmed.

Finally, let us have a brief recall on what we have accomplished in this chapter. Most succinctly, this chapter has gone along the following route. Length contraction and time dilation (which are the two <u>core</u> concepts of special relativity) are factually incompatible thus essentially contradictory with each other (a plain fact) → these two <u>core</u> concepts are clearly and seriously self-contradictory (a solid fact) → special relativity turns out to be clearly and seriously self-contradictory (an unavoidable fact) → this unavoidable fact has been verified by revealing and showing that the famous Twin Paradox turns out to be clearly, even obviously, a paradox → this unavoidable fact has been confirmed by revealing and showing that the five <u>main</u> components of special relativity cannot be put into the same package of special relativity at all; this confirmation is a further demonstration of the self-contradictory feature of special relativity from the wide angle of its entire structure. As a result, the above route clearly indicates that the take-home message of this chapter is: special relativity turns out to be a theory that is clearly and seriously self-contradictory, being the gravest, also extremely disastrous, consequence of the basic fact that special relativity doesn't have the ability to know *why* the speed of light is constant.

ABOUT THE AUTHOR

Bingcheng Zhao, who was born in 1963 in Shandong Province of China, obtained his Ph.D. in 2001 from Washington State University. He is the author of the popular science books: *Why It's Difficult to Understand "A Brief History of Time"*, published in 2016; and *Dark Matter Is No Longer in the Darkness! (The Constituents of Dark Matter Have Been Revealed)*, published in 2018. He is also the author of the academic book: *From Postulate-Based Modern Physics to Mechanism-Revealed Physics*, published in 2009; the newly developed and verified mechanism-revealed physics is the key to solving the fundamentally important problems of dark matter and dark energy; and the birth of mechanism-revealed physics actually heralds that the spring of science is coming again.

www.ingramcontent.com/pod-product-compliance
Lightning Source LLC
Chambersburg PA
CBHW030952240526
45463CB00016B/2516